PEOPLE AND THEIR PLANET

Also by Barbara Sundberg Baudot

INTERNATIONAL ADVERTISING HANDBOOK: A User's Guide to
Rules and Regulations

People and their Planet

Searching for Balance

Edited by

Barbara Sundberg Baudot
Professor of Politics
Saint Anselm College
Manchester, New Hampshire, and
Research Associate, Global Development and Environment Institute
Fletcher School of Law and Diplomacy, Tufts University
Medford, Massachusetts

and

William R. Moomaw
Professor of International Environmental Policy, and
Director, Global Development and Environment Institute
Fletcher School of Law and Diplomacy, Tufts University
Medford, Massachusetts

Foreword by Nafis Sadik

First published in Great Britain 1999 by
MACMILLAN PRESS LTD
Houndmills, Basingstoke, Hampshire RG21 6XS and London
Companies and representatives throughout the world

A catalogue record for this book is available from the British Library.

ISBN 0–333–68811–2

 First published in the United States of America 1999 by
ST. MARTIN'S PRESS, INC.,
Scholarly and Reference Division,
175 Fifth Avenue, New York, N.Y. 10010

ISBN 0–312–21715–3

Library of Congress Cataloging-in-Publication Data
People and their planet : searching for balance / edited by Barbara
Sundberg Baudot and William R. Moomaw : foreword by Nafis Sadik.
p. cm.
Includes bibliographical references and index.
ISBN 0–312–21715–3 (cloth)
1. Population—Environmental aspects. 2. Environmental policy.
3. Population geography. 4. Human geography. I. Baudot, Barbara
Sundberg. II. Moomaw, William, 1938– .
HB849.415.P44 1998
304.2—DC21   98–23533
                                                        CIP

This book is printed on paper suitable for recycling and made from fully managed and
sustained forest sources.

10   9   8   7   6   5   4   3   2   1
08   07   06   05   04   03   02   01   00   99

Printed and bound in Great Britain by Antony Rowe Ltd, Chippenham, Wiltshire

*First published as 'Water Conflicts along the U.S.–Mexico Border: Towards a Transboundary Water
Market?' in *The Scarcity of Water: International, European and National Legal Aspects*, eds E. J. de
Hann and E. Brans (Kluwer, 1997).

To **Nancy Anderson**

whose most cherished life purpose it was to encourage all people to make environmental activism a part of their lives and who prayed that 'the 21st century will truly find that the frightening uncontrolled greed of the past is stopped. This must be the earth's century and humankind must harmonize with her.'

Contents

Contents   ix

List of Tables

List of Figures

Acknowledgments

The initial ideas for this book were gathered at a Conference on Population and the Environment, convened at the Fletcher School of Law and Diplomacy in June 1994. Subsequently, a number of the participants and other scholars, pursuing their academic or other research interests on the relations between population and environment, agreed to contribute the papers for this collection.

The editors gratefully acknowledge support for the production of this book provided by the Global Development and Environment Institute (GDAE) of the Fletcher School of Law and Diplomacy, Tufts University, the Kendall Foundation, and the United Nations Population Fund (UNFPA).

The greatest debt of gratitude is owed to Carolyn Logan, our technical editor, whose tireless efforts in organizing, correcting and unifying texts have been essential to bringing this work to light.

We are also grateful to Al Belote for his genius in making ready all the figures and tables. Renée Bleau and Robin D'Alessandro deserve credit for their considerable secretarial assistance. Thanks are also due to Elise Laura, Amelie and Jacques Baudot, as well as Ruth Sundberg, for their willing help.

Finally, we are particularly appreciative of the wonderful collaboration we have had with a group of very patient, cooperative and understanding contributors, all of whom are working for a more sustainable environment.

BARBARA SUNDBERG BAUDOT
WILLIAM R. MOOMAW

Notes on the Contributors

Virginia Deane Abernethy is Professor, Department of Psychiatry, Vanderbilt University, Nashville, Tennessee, USA.

Gottfried Tenkorang Agyepong, with **Edwin A. Gyasi, John S. Nabila** and **Sosthenes K. Kufogbe**, works in the Department of Geography and Resource Development, University of Ghana, Legon-Accra, Ghana.

Barbara Sundberg Baudot is Professor (Political Science) Saint Anselm College; Research Associate, Global Development and Environment Institute, Fletcher School of Law and Diplomacy, Tufts University, Medford, Massachusetts, USA.

Richard E. Bilsborrow is Research Associate Professor, Department of Biostatics as Fellow, Carolina Population Center, University of North Carolina, Chapel Hill, North Carolina, USA.

John Elder is Professor, Program in Environmental Studies, Middlebury College, Middlebury, Vermont, USA.

Robert Engelman is Director, Population and Environment Program, Population Action International, Washington, DC, USA.

Nasim Firdaus is Director, Ministry of Foreign Affairs, Dhaka, Bangladesh.

Laurel Heydir works in the School of Law and Population Research Center, Sriwijaya University, Palembang, Indonesia.

Haydea Izazola is a researcher at the El Colegio Mexiquenza, Mexico.

Jeffrey N. Jordan is Deputy Director the Futures Group International, Washington, DC, USA.

Sai Felicia Krishna-Hensel is Coordinator, Interdisciplinary Global Studies Research Program, Center for Business and Economic Development, Auburn University, Montgomery, Alabama, USA.

Véronique Lassailly-Jacob is Research Fellow, Centre for African Studies, French National Council for Scientific Research (CNRS), Paris, France.

Catherine M. Marquette is a consultant at the Michelsen Institute, Bergen, Norway.

William R. Moomaw is Professor, Co-director, Global Development and Environment Institute, Fletcher School of Law and Diplomacy, Tufts University, Medford, Massachusetts, USA.

Francisco Pichón is affiliated with the World Bank and the University of North Carolina at Chapel Hill.

Dennis Pirages is Professor, Department of Political Science, University of Maryland, College Park, Maryland, USA.

Sandra L. Postel is Director of the Global Water Policy Project, Amherst, Massachusetts, USA.

Roberto A. Sánchez-Rodriguez is Acting Associate Professor in Environmental Studies, University of California, Santa Cruz.

Alex de Sherbinin is a University of Michigan Population-Environment Fellow at IUCN – the World Conservation Union, Gland, Switzerland.

D. Mark Tullis works for Folio Inc., Provo, Utah, USA.

Foreword

Policy-makers the world over are now more than ever aware of the importance of considering the population factors in all issues relating to sustainable development. A good basis for the awareness of the complex linkages between population, environment and development was established at the United Nations Conference on Environment and Development (UNCED) in 1992. As former Norwegian Prime Minister, Gro Harlem Brundtland, noted, 'poverty, environment and population can no longer be dealt with – even thought of – as separate issues'. This conviction is reflected in Agenda 21, which concludes that 'the growth of world population and production, combined with unsustainable consumption patterns, places increasingly severe stress on the life-supporting capacities of our planet'.

The unprecedented growth in human numbers in our time will have profound effects on our physical environment. Population is growing faster than ever before, at 93 to 95 million people a year. Our current population of 5.6 billion will grow to 6.2 billion by the end of the century. Nearly all of this growth will be in Asia, Africa and Latin America. Over half will be in south Asia and Africa, the poorest regions in the world. One result of this skewed growth is to stimulate movement. People are leaving rural areas for the cities in greater numbers than ever; by the end of the century, half of the world's population will live in cities. Conversely, some of the world's poorest people are invading forests and fragile watersheds in search of land and livelihood.

Population growth also means that there are more very poor people in the world than ever before, and there are fewer prospects for an improvement in their lives. If the world is to make progress towards development and avoid the destructive effects of poverty, some way must be found to give hope to the 'bottom billion' poorest people. Some means must also be found to meet the aspirations of the three billion people who are neither very poor, nor very affluent.

At the same time, we must not forget the responsibility of the developed world. In the developed world, industrialization, urbanization, chemical-intensive farming techniques and wasteful consumption patterns, among other factors, interact with population

growth to cause substantial environmental degradation, both globally and in the developed countries themselves. All of these changes
will have deep and interacting environmental and developmental
effects. The challenge is to raise the living standards of four-fifths
of the world's people, an increasing number of whom live in cities
and who are increasingly mobile, without destroying the environment on which we all depend. We need development; but development without sustainability defeats its own purpose.

There is reason for optimism, however, because a new vision and
approach to development is taking shape. It is well-embodied in
Agenda 21, and it has been further carried forward in the International Conference on Population and Development (ICPD). The
Cairo Conference came at a pivotal time, between the United Nations
Conference on Environment and Development (UNCED) in 1992,
and the Social Summit (WSSD) and the Women's Conference, both
held in 1995. It was the first international conference convened by
the UN explicitly to address population concerns in terms of their
inter-relationships with sustainable economic growth, sustainable
development, poverty alleviation, gender equality and reproductive
health.

Most importantly, the ICPD, the third UN Conference to focus
on population, was the last major international opportunity in this
century to consider past progress and what still needs to be done
in the area of population. Alongside the fine stirring speeches, the
declarations of commitment and the expressions of hope, there is
the major outcome of the Conference: an international Programme
of Action on population and development, adopted by 179 delega-
tions, and embraced by over 1000 NGOs. The ICPD goals were
adopted from other international goals that had already been agreed
upon, such as the Alma Alta goals for Health for All, the Jomtein
Education for All goals, the World Summit for Children, and
UNCED. As such, they support coherent and coordinated action
throughout the UN system and the world community of nations.

At the heart of the Programme of Action is the recognition that
population issues are an integral part of, and should be at the centre
of, our common efforts towards sustainable development. Population must be fully integrated into, and become a central component of, the development process if the efforts of individuals, nations
and the international community are to bring about equitable, hu-
man-centered sustainable development. A key concern of the
Programme of Action of ICPD is the recognition that the empower-

ment of women is essential for both individual and broader developmental goals, and that women's education, protection from discrimination and violence, and equitable participation in the policy-making process should be ensured.

Another significant aspect of the Programme of Action is the call for a renewed commitment on the part of developing countries, as well as donors, to provide the necessary financial resources to achieve population and developmental goals. The international donor community has reiterated its growing concern about population issues world-wide on several occasions. These concerns have not always been translated into adequate funding of population activities. Today only about 1 per cent of official development assistance is for the population sector.

Social and environmental change, including population change, is taking place on a scale never before witnessed. This change may both add to and detract from such stability as exists today. There are bound to be tensions and adjustments. In this context, the international conferences – including UNCED and ICPD, as well as the Social Summit and the Women's Conference – have an extremely important function in terms of consensus building. Likewise, they help maintain and reshape unity. As such, they contribute to our common security.

In summary, there are perhaps two overriding themes that stand out in the ICPD process. One is the full integration of population concerns into development; the other is the centrality of human beings and the respect for individual choice. We must not allow these themes to be diminished, watered down, or obscured.

NAFIS SADIK
Executive Director
United Nations Population Fund UNFPA

Overview: The Population–Environment Equation: Implications for Future Security

Barbara Sundberg Baudot

Sustainable development implies an enduring balance between the numbers of people inhabiting the planet and the organic and inorganic resources that support human life. This balance can be represented metaphorically as a complex mathematical equation that relates population and environmental factors, both at an aggregate, planetary level and at a disaggregated level based on geographic eco-regions and political divisions. The factors in this equation are derived from conceptions of the environment as living space and source of basic resources, as the primary life-support system, and as an aggregate of social and cultural conditions that influence the lives of individuals and communities. Population factors reflect demographic trends and a broad spectrum of mentalities and behavioural phenomena that encourage or discourage population expansion and the consumption of resources, and determine the overall treatment of the environment. These factors are conditioned by a number of variables associated with diverse cultures, education levels, and value systems that determine life-styles, design development paths, and fuel scientific and technological progress.

The combination of exponential population growth, increasing scarcity of resources and living space, and unbridled economic growth and consumerism in the world today suggest that the equation is out-of-balance. This imbalance may diminish the capacity of the planet or of specific geographic regions to provide sustainable habitats and life-supporting resources and to preserve biodiversity. The resulting environmental insecurity is reflected in individual fear, anxiety in anticipation of violence, and socioeconomic injustices that are causally linked to these environmental transformations. Reactions range from acceptance and adaptation, to conflict, depending on the particular ecosystems, the prevailing social ethos and the effectiveness of political institutions.

Rebalancing and stabilizing the population–environment equation requires modifying the life-styles and expectations of millions of people. Political measures, research, education and moral persuasion are some of the steps that can be taken as societies seek to reach a balance. But the necessary political and private responses will not usually occur before credible evidence of an imperiled environment convinces decision-makers that change is essential.

This book brings together research that has been carried out on the relations among the factors in the population–environment equation, on the symptoms of imbalance, and on ways to establish a balance. The research examines several important but as yet only partially understood factors, as well as under-scrutinized cause–effect relations, and subjects that beg to be revisited from multidisciplinary viewpoints. The book is divided into three parts. The first focuses on the factors in the equation and the relations between them, the second examines evidence that the equation is out-of-balance, and the third considers the search for balance.

PART I: THE FACTORS IN THE EQUATION

Analysis of the relationships between the factors in the equation is the prelude to determining the causes of environmental change and the role of population pressure. There are two dimensions to the research contained in this section, the first focusing on developing methodologies, and the second on field studies of resource–population relations.

Ensuring that population, environment and poverty factors are integrated into sustainable development policies challenges the capacities of researchers and scholars to develop reliable measures of these elements and their interlinkages. Although it would be useful to be able to capture the marginal environmental changes that are directly related to population growth, isolating the effects of this demographic factor raises profound conceptual and methodological questions. Environmental changes cannot be attributed solely to demographic factors because demographic pressures are inextricably connected with consumption, technology, social values and political will. Given the labyrinthine nature of the links between demographics, production processes, patterns of consumption and other aspects of human behaviour, a holistic approach is taken using models that advance understanding of the complexities of population–environment linkages.

Although many of the studies in this volume apply specific methods of analysis, three papers essentially focus on methods for building and assessing relations between population and environmental factors. Catherine M. Marquette and Richard E. Bilsborrow comparatively survey the methodologies that are presently used in the study of population impacts on the environment. Their survey reviews the principle methodologies applied in the 1990s, evaluates the strengths and weaknesses of each as they have been revealed in selected research projects in developing countries, and highlights areas for future research. Jeffrey N. Jordan proposes a conceptual framework for analyzing the relationship between population and the environment, and applies it to policy development that addresses desertification in Africa, degradation in the Bay of Bengal, and the problems of small island states. William R. Moomaw and D. Mark Tullis apply the IPAT equation in a comparative study of the causal factors that contribute to climate change in 12 countries at different stages in their developmental trajectories. The study measures the relative contribution of population, affluence and technology to the production of CO_2 emissions.

The geophysical repercussions of population growth will vary from place to place depending upon the fragility of the resource base, population densities and the prevailing societal ethos. Understanding these varying interrelationships can be advanced by collecting case studies and by building databases on demographic trends and relevant socioeconomic factors. Two case studies and two global studies focusing on particular resources illustrate this approach. Francisco J. Pichón's case study investigates the factors influencing land use and deforestation by migrant colonists in the Amazonia region of Ecuador. The study reveals the vulnerability of migrants to the dictates of a fragile resource base. Laurel Heydir describes factors affecting the deforestation of the Lahat region in South Sumatra Province of Indonesia. He demonstrates that a combination of increasing population, growing demand for arable land, and inadequate management by the government has resulted in extensive deforestation.

Sandra L. Postel describes the linkages between increasing water scarcity and growing populations on a global scale, and also offers comparative country data. She presents water scarcity as a rapidly emerging constraint to raising living standards and meeting basic needs in many regions, and determines that present consumption levels are unsustainable. Robert Engelman describes the implications of population as a scale factor, and applies this to the availability of three critical resources: cropland, water and fish.

PART II: THE EQUATION OUT-OF-BALANCE

Reactions to changes in the physical and social environment range from increased optimism and fertility, to fear, violence and widespread conflict, in a spectrum that also includes mounting indifference, increasing social malaise, declining health, increasing socioeconomic injustice, and migration. These responses can, in turn, exacerbate the impact of population pressure on the environment. The equation is out-of-balance when population pressures and behaviour operate to overcome the carrying capacity of a particular ecosystem.

The determination that environmental changes actually diminish carrying capacity and threaten human security shifts inquiry to the study of less tangible phenomena. A threat is a feeling or an intuition of risk, peril, menace, danger or fear – the prelude to insecurity to be experienced by individuals and, by extension, whole societies. But events or conditions traverse belief thresholds in various ways as they inflict fear and raise concern. At times, the effects of conditions that most reasonable people would regard as threats are mitigated by circumstances that instead entice a populace to ignore or to sublimate its normal apprehension. Research has, however, identified environmental conditions that cause most societies to experience malaise and social injustice. They include urban squalor, high pollution levels, perceived losses of life-sustaining resources, unemployment, declining quality of life, and forced changes in life-style.

Several studies presented here treat different aspects of the *equation out-of-balance*. Virginia Deane Abernethy points to evidence that it is environmental conditions that encourage or discourage population growth, rather than development as many scholars have postulated. This thesis parallels the classic observations of nature and animal behaviour described by David Lack (1954) in *The Natural Regulation of Animal Numbers*.[1] When people experience no environmental constraints, they are encouraged to have many offspring; the reverse is true when there is a sense of deprivation.

One of the main symptoms of overpopulation is the emergence of megacities. This phenomenon and the problems associated with it in India are described by Sai Felicia Krishna-Hensel. Haydea Izazola and Catherine M. Marquette address the problems of Mexico City, another of the world's megacities, charting both the out-migration of the middle class as pollution and population crowding increase, and the resultant exacerbation of environmental problems as the poor are increasingly left to cope with the growing burden of degradation.

Over the past decade the world refugee population has more than doubled from 8 to 18 million. Driven by environmental stress and other political and economic factors, migrants put pressures on host land with their demands for food, jobs and living space. The net impact of environmental refugees is more environmental stress and political tension. Véronique Lassailly-Jacob describes conflicts in refugee camps between migrants and indigenous inhabitants in Zambia, and argues that the government and the international community have not been able to deal adequately with these problems.

Political unrest and even armed conflicts can accompany population pressures and scarce resources unless appropriate measures are taken. Roberto A. Sánchez describes the potential for conflict between Mexico and the US over scarce water resources along the border between the two countries, but also the ways in which the two countries have managed to avert any outbursts, in part through the development of informal, transboundary water markets.

PART III: SEARCHING FOR BALANCE

When disequilibria exist, measures must be taken to correct the imbalance, but formulating policy entails extensive negotiations and often acrimonious debate. Before any adverse environmental condition can be remedied, people must recognize that the problem bears sufficiently high negative social costs to warrant the investments of time, money and resources required to overcome it. Yet many people will not attribute high negative social costs to changes that occur in faraway ecosystems that are not readily perceived, nor to imbalances that develop gradually allowing subconscious adjustment to the altered conditions. What influences create perceptions that deteriorating environmental conditions imperil human security and that action must be taken?

One key question is what motivates people to have large families. Nasim Firdaus suggests that gender bias, financial insecurity, labour demand, and cultural and humanist values play important roles in determining family size. Achieving balance therefore depends on giving attention to all these dimensions. Alex de Sherbinin pinpoints consumerism and the trade activities of the affluent as major causes of disequilibrium, and argues that balance requires gaining control over excessive materialism. The case study of the fragile ecosystem of the Western Saharan savanna by G.T. Agyepong, E.A. Gyasi, J.S. Nabila and S.K. Kufogbe describes a plan for sustainable development that builds on the cooperation and culture of the people involved. John Elder's work demonstrates how literature, and especially poetry, can be used to inspire efforts to achieve a balance. Dennis Pirages discusses ecological security and restoration of balance from the micro-perspective, which involves a major reassessment of the general direction of sociocultural evolution.

The following introductory chapter by Barbara Baudot provides an overview of the three themes of this book recast as the physical, social and personal dimensions of a potentially sustainable environment. Balance depends on simultaneous efforts on all three planes. Policies are required that deal with the physical manifestations of imbalance, but these efforts are in vain if, on the social plane, economic growth and industrialization remain the principle goals of modernization and development. And, in the end, the personal dimension is primordial. Balance begins with each individual's compelling concern for his or her environment, from the microcosm of the personal mind-set, to the macrocosm of the planetary universe. The substance of law depends on what people value.

Note

1. David Lack, *The Natural Regulation of Animal Numbers* (London: Oxford University Press, 1954).

1 Introduction: Dimensions of the Population–Environment Equation

Barbara Sundberg Baudot

On the road to the year 2000 the world confronts many challenges. The environment is rapidly deteriorating, and poverty and frustration are endemic and spreading. Civil strife is commonplace. At the same time population growth has escalated, increasing ten-fold since 1750, two-fold since 1950. There is concern that the globalization of acquisitiveness, materialism and individual freedom without personal and social responsibility will irrevocably upset the fragile balance that exists between human society and the natural environment. Such problems have been drawing increasing attention from scientists, philosophers and society, particularly since the 1960s, although these issues have been concerns since the time of Malthus. They give rise to a number of key questions. What are the dimensions of the population–environment equation? What must be understood about the problems on each of the dimensions to address disequilibria effectively? How can a balance between human activity and environmental factors best be achieved that will permit human flourishing and maintain the integrity of the planet? Before addressing these questions, a few initial clarifications are necessary.

Several years ago, Lynton Caldwell wrote that mankind lives in two realities: the abiding reality of the earth – that is, the planet, independent of humans and their works – and the transient reality of the world, a product of perception and imagination:

> The earth and its biosphere form a grand synthesis of complex interactive systems within systems ... The world is the way humanity understands and has organized its occupancy of the earth: an expression of imagination and purpose materialized ... Oceans, islands, species, and ecosystems are integral parts of the earth, but the world is not integrated – its cultures and their [respective]

1

values do not comprise a unity. Physically [people] belong to the earth, . . . intellectually they may transcend it . . . The so-called environmental crisis [today] derives from this physical and intellectual duality .[1]

In considering Caldwell's vision, it is legitimate to ask whether the human intellect is capable of perceiving the *abiding reality of the earth*. Through studying changes in physical phenomena, scientists are able to surmise the probable consequences of human activity on defined physical surroundings, and they can make less-certain assumptions about the effects on the biosphere. But determining such factors as the carrying capacity of the planet remains problematic. Moreover, although they are empirically founded, scientific theories remain in the category of world views. Albert Einstein, for example, stated that:

> The whole of science is nothing more than a refinement of everyday thinking . . . But even the concept of the 'real external world' of everyday thinking rests exclusively on sense impressions . . . the differentiation between sense impressions and images is not possible: or, at least it is not possible with absolute certainty.[2]

By this statement and many similar ones, Einstein implies that scientific theories are fundamentally creations of the human mind, and while based on observation and accumulated knowledge, they remain no more than world views. Louis Michel, French mathematician and physicist, finds that the noblest attribute of scientists is humility, the ability to accept being confronted with many facts that defy complete understanding.[3]

Policy-makers recognize the limitations of human knowledge. Because comprehensive environmental studies are fraught with scientific reservations, the *precautionary principle* has evolved as *the next best* justification for international treaties and national policies aimed at protecting the environment. This principle suggests that it is better to err on the side of prudence and to take measures to protect the environment in the present in order to avoid potentially irreversible catastrophes in the future. One prominent example of the application of this principle is the Convention on Climate Change.

In light of the above, the human–environment relationship can best be depicted as a collage composed of diverse concepts, beliefs and theories that emerge from various perceptions of the illusive realities of the earth. The challenge is to derive from this assemblage the sustainable population–environment balance that is most likely to secure a satisfying life experience for generations to come. One way to address this challenge is to consider the overall equation from the perspectives of the three principal levels of human interaction with the environment – the physical, the social and the personal – and then to argue that achieving a sustainable equilibrium depends on realizing consistent balances on each plane. The three dimensions can be defined as follows:

- The **physical** – on the surface of life, the population–environmental equilibrium is a function of the balance between the number of people and the resources and space necessary to sustain a minimal quality of existence.
- The **social** – this is a construct of ideologies and social behaviours that build and shape communities, develop social and economic conditions, and transform properties of nature into artifacts, habitats and urban landscapes. Balance depends on the conceptions and practices of development and progress, whether they emphasize material progress or human flourishing through advances in the art of living. Disequilibria are evident in civil strife and gross human inequities. Disequilibria are inevitable where the market determines human success and circumscribes social relations.
- The **personal** – this is the dimension of the human mind. Specifically, the contest in this realm is between the drive to control and consume nature's bounty, juxtaposed with the longing to exist in harmony with nature's laws. Disequilibria occur when materialism and anthropocentrism govern individual thought.

Achieving overall balance begins on the personal level, where beliefs, values, and interests create motivations, and passions emerge. These ideas and forces motivate and direct action on the social plane, where impacts are discerned and then treated on the physical plane.

The following sections develop these three perspectives, beginning with the physical, and ending with the personal. Although cause–effect logic would argue reverse treatment, in reality it is the physical plane that gets the most attention. The intent is to move from the material to the more intangible – but proportionately more significant – factors in the equation.

THE PHYSICAL DIMENSION

This is the immediate aspect of the population/environment equation. The environment is a composite of the material surroundings, resource bases and physical habitats of humankind. Independent variables include demographic trends, patterns of production, consumption and resource renewal, and technological adjustments. Disequilibria occur when and where population crowding causes people to suffer from lack of vital resources and pollution, and thus to lose their sense of well-being and security. Migrations, conflicts and diseases ensue. Deforestation, desertification, pollution, biospheric change and declining biodiversity are indicative of disequilibria.

The principle variables on this dimension are generally quantifiable. However, despite the tangibility of the evidence, the statistical picture is far from complete. The earth is a mosaic of overcrowded and underpopulated regions, with such a variety of resource endowments that a great deal more research is required to ascertain the full physical dimensions of the population–environment equation.

It is euphemistic to suggest that the dominant political ethos in today's globalized free enterprise economy makes it difficult to achieve balance in the physical dimension, especially when rugged individualism and consumerism are considered vital to the progress of the world economy, and scientific uncertainty prevails about the impacts of demographic trends and environmental changes. The extreme difficulty of making objective political decisions has been observed by economist Richard Mancke in his book, *The Failure of US Energy Policy*.[4] The problems of decision-making begin with the political nature and subjectivity of the concept of *pollution*. According to Mancke, an economic activity is *polluting* whenever society's evaluation of the total costs attributable to it, including the costs of its undesirable by-products, exceeds society's evaluation of its total benefits. In order to measure social costs and benefits, every society must decide whose world views should be considered legitimate and how heavily these different perceptions are to be weighed. Since there are no objective tests that enable governments to make these decisions, the answers depend on political processes and subjective value judgments. While disputes are inevitable, ultimately the views of the most persuasive and influential groups prevail. Moreover, consistent with such political/economic realism, it is unlikely that any moral philosophy or set of values can guide the selection of measures designed to deal with imbalances. Thus, in such a context, fear becomes the strongest motivator for action, and crisis mentalities the most effective forces for change.

Underlying the debates on population crowding and environmental stress lies a jumble of unexplored – but not irrelevant – considerations, beliefs and values that express people's aspirations and needs. Failure to take these into account can spell doom for the most carefully designed, but scientifically-based, policy. Thus, balance in the physical dimension of the population–environment equation requires consideration of related issues on the social and personal planes. In a democratic society, the individual sense of social responsibility must balance the desire for personal gain, in order to bring about successful implementation of environmental policies, particularly when voluntary cooperation is needed and enforcement is problematic.

THE SOCIAL DIMENSION

The global society is a vast interplay of more than five billion personalities, each with different histories, beliefs, ambitions, tastes and loves. Gathering in diverse communities and institutions, they collectively define the social ambiance and conditions of human life. It is people who fuel social progress, create and consume wealth, develop science and technology, and, by their activities and interactions, shape and transform the environment. Thus, the dynamics on the social plane are critical to environmental security. On this level, the challenge is to establish a balance between the competitive spirit and the drive for material progress on the one hand, and, on the other, the forces that favour sustainability, including longings to exist in harmonious, mutually benefiting and humane environments where progress is measured by the enrichment of the life experience.

The key question for equilibrium on this plane is: what constitutes a developed society? While development is usually associated with a vision of progress from a primordial antediluvian state toward the full flourishing of human potential, this view leaves room for very different conceptions of development. The prevailing concept is *modernization*, broadly defined as the process of increasing knowledge of technology whereby individuals gain control over their environment while progress is measured by the rate of economic growth and the increase in material well-being. A much less common vision emphasizes the process through which people gain security and stability, and exercise their freedom of action within the constraints of wisdom and custom and in close interaction with others. In this view, progress is measured in terms of cultural and spiritual advancement and an increased capacity to practice the art of living. To this end, improved economic conditions are a means to remedy causes of human suffering, but not the *sine qua non* for, nor the evidence of, growth and development.[5] These contrasting concepts point towards alternative paths to the future. Obviously, the second concept is more consistent with aspirations for achieving balance in population–environment equations.

The course of *modernization* is directed towards upward trends in economic productivity, applied science and technology, and GNP per capita. It is most easily travelled by nations with competitive cultures wherein the exercise of individual freedom is generally free from government-imposed social responsibilities, and the pre-eminent qualifications for success or survival are acquisitiveness and market prowess. Modernization advances a concept of human dignity that is contingent upon increments of wealth, power and respect for human rights. The natural environment fuels the economic engine by providing land for construction, sinks to absorb wastes, and, with what remains, offers a refuge for peace of mind and recreation.

The fundamental axioms of modernization derive from western enlightenment theory, especially the economic and political thinking of Adam Smith and John Locke, and their contemporaries on the European continent. The enlightenment lifted human thought above the oppressive medieval dictates of superstition and irrational mythology. People became sovereign over their own progress and destiny. Based on such assumptions, Adam Smith laid out the first plan for economic development, whereby governments are to ensure the conditions that will allow people to follow their natural propensities 'to truck, barter, and exchange one thing for another.'[6] Smith argued that in human affairs, progress is the ruling tendency, and his plan was based on his confidence in the innate self-interest of human beings and the natural harmony of these interests. According to Smith:

> The uniform, constant, and uninterrupted effort of every man to better his own condition, the principle from which public and national, as well as private opulence is originally derived, is fre-

quently powerful enough to maintain the natural progress of things toward improvement.[7]

Smith envisaged people operating in their own economic self-interest,

> ... led by an invisible hand to promote an end which was no part of [their] intention. Nor is it always worse for the society that it was no part of it. By pursuing [their] own interest [people] frequently promote that of the society more effectually than when [they] really intend to promote it.[8]

Smith's economic theory found its complement in the political theories of John Locke, who argued that individuals must live freely and harmoniously in societies, establishing governments to protect their lives, liberty and property.[9]

Smith's and Locke's theories were underpinned by ethical assumptions rooted in an historical context with many social, moral and religious constraints. The ultimate objective of these English liberal theorists was the common good, an equitable and fair society. For Smith, the invisible hand would work to this end under conditions of fair play, including free competition, unobstructed information flows, and equal access to the market. He also assumed that the supply of resources would be more than ample. Thus, in the market, human beings would pursue their self-interest in ways that violated no ethical canons.

Turning Locke's theory of property into the classic doctrine of the *spirit of capitalism* can be done only by explaining away all of the statements that he made about the origin and limitations of property.[10] Locke considered property a means to an improved human race, implying that the accumulation of property was important in so far as it would free people to perfect their characters and human qualities. The society was to be governed according to the same justice that prevailed in the state of nature, that is according to the Golden Rule and other principles of virtuous behaviour included in the supreme Law of Nature.

Today, the threats of serious imbalance in the social dimension are many and varied. Ultimately, they can impede the best efforts to ensure sustainable population–environment relations. Some of the more salient pitfalls on the modernization path are:

- loss of the ethical basis for the *laissez-faire* doctrine;
- diminishing opportunities for expression of human ingenuity and creativity;

- weakening of individual responsibility; and
- emergence of a market society.

The ethical underpinnings of the modernization path are destroyed when power seekers transform benign theories into ideologies and, in the process, adopt only those prescriptions that advance individual wealth and power. For example, when democratic capitalism depends on the invisible hand of rugged individualism in a *laissez-faire* market of imperfect competition, social Darwinism and scarce resources, it is highly improbable that the sum of 'individual trucking, bartering, and exchanging' in the marketplace will bring about the provision of public goods, internalize the costs of externalities in the production processes that blight the environment, or provide for the well-being of the individuals who are excluded from the market as cost-efficient, labour-saving machinery and electronic devices render them obsolete. Politically, unquestioned faith in self-interest coupled with support for the virtues of aggressive egoism would seem to threaten the good works of progress, reason and industry.

Smith himself recognized that the marketplace alone would, in due course, be unable to rectify the inevitable negative impacts of uncontrolled commercial spirit. Noting that progress would lead to the division of labour and specialization, he argued that the government must protect society from the inevitable social costs of progress, including the widespread degeneracy, vacuity and ignorance that accompany unbridled technological progress. He stated that:

> In the progress of the division of labour, the employment of . . . the great body of the people comes to be confined to a very few operations . . . The man whose whole life is spent in performing a few simple operations . . . has no occasion to exert his understanding, or to exercise his invention in finding out expedients for removing difficulties which never occur. He naturally loses, . . . the habit of such exertion, and generally becomes as stupid and ignorant as it is possible for a human creature to become. The torpor of his mind renders him, not only incapable of relishing or bearing a part in any rational conversation, but of conceiving any generous, noble, or tender sentiment, and consequentially of forming any just judgment concerning many even of the ordinary duties of private life . . . His dexterity at his own particular trade seems, . . . to be acquired at the expense of his intellectual, social, and martial virtues. But in every improved and civilized

society this is the state into which ... the great body of people, must necessarily fall, unless government takes some pains to prevent it.[11]

The implications of Smith's observations bode ill for the natural environment, which pales in the distant background compared to the market's immediate and tangible concerns. Little can be expected from people buried in the tedium of mechanistic functions in the absence of efforts on all levels to ensure cultural growth and general stimulation of the human spirit.

Another pitfall in the modernization model is the gradual disappearance of the notion of social responsibility. This process has been explored by philosopher Tomonobu Imamichi, who argues that responsibility is ontologically tied to the individual. But it is problematic to view ethics as the self-realization of the spirit when humanity is confronted with the moral dilemmas of the technological society. The goal-orientation of human thought that made scientific development possible has also moved society from a man-controlled, instrumental technology to an age of autonomous scientific and technological power. In this age, moral decisions tend to escape individual control, falling under the anonymous responsibility of the multiple owners of technology. When the group, the committee, the board or society at large becomes the arbiter of moral decisions, responsibility has lost its ontological roots and must be re-established according to a collective eco-ethic of responsibility to ensure equilibrium within society and between society and the environment.[12]

There is also the danger that societies will become obsessed with money and materialism to the point that there are few opportunities for human relations that are not confined to financial, market or technological channels. Karl Marx described a similar phenomenon as *commodity fetishism,* whereby significant relationships between people in society become functions of the respective products each offers in market exchange.[13] The modern manifestation of this phenomenon is the *market society*. In contrast with a *market economy*, which promotes individual initiative and exchange of goods and services and provides a reasonable allocation of resources among members of the community, a market society extends economic rationality to all spheres of life and to all facets of human and social relations.

The predisposition of democratic societies for the development of market societies and for a proportional loss of mutually-caring communities was identified by Alexis de Tocqueville almost 200 years ago. He observed that lurking beneath the great benefits of social equality were propensities for people to distinguish themselves from others, to isolate themselves from one another, to be self-absorbed, and to become inordinately in need of material gratifications. He also observed that, when the taste for material gratification or commodities outruns the constraints imposed by peoples' education and sense of political responsibility, communities will decay:

> The time will come when men are carried away and lose all self-restraint at the sight of the new possessions they are about to obtain. In their intense and exclusive anxiety to make a fortune they lose sight of the close connection that exists between the private fortune of each and the prosperity of all.[14]

Nigerian social scientist Claude Ake identified similar consequences of the market obsession. While noting the great benefits of a free market, Ake found that 'the forces that underlie the success of the market – egotism, individualism, self-interest – are also responsible for undermining society's strength which lies in a sense of community.'[15]

Not unrelated to the reification of social relations in the market society is the marketing of sexual permissiveness. Sexual allure has always been exploited in the underground markets of prostitution, but in a market society it can become a particularly potent stimulant of demand. It is *the way* to sell cigarettes, automobiles, high-fashion jeans, and just about everything else. Market eroticism has global reach through satellite communications, business, and trade. When sexual relations are promoted and widely counselled as a *product* and a *performance*, the most serious threat to social balance may become the trivialization, fragmentation and depersonalization of human life. It is paradoxical that today individuals and societies are asked to limit their procreation while the market surrounding them encourages libidinous behaviour.

The market society creates a dichotomy between the appearance of sociability and compassion, and the cold reality of calculation, alienation and atomism. Concern for preserving the natural environment plays little role in its ethos, which makes instrumental or utilitarian use of its natural resources. The few efforts to conserve the environment are buried in such thorny issues as the financial implications of technology transfer, the exploitation of patent rights and know-how, and a myriad of other transactions relating to money and power. Environmental insecurity is rarely factored into the *conscious* of the market society unless human health is in jeopardy, business deems it prudent, or producers are under pressure to do so from government or civil society.

Where does the modernization path end? One answer is unwittingly, but dramatically, offered by Roland Barthes in his critique of classical Dutch painting, wherein all vestiges of the sacredness of life have faded away, to be replaced with humankind and its empire of things:

> [Humankind] stands now, [their] feet upon the thousand objects of everyday life, triumphantly surrounded by their functions. Behold [them], then at the pinnacle of history, knowing no other fate than a gradual appropriation of matter. No limits to this humanization, and above all, no horizon: in the great Dutch seascapes (Cappella's or Van de Venne's), the ships are crammed with people or cargo, the water is a ground you could walk on, the sea completely urbanized ... As if the destiny of the Dutch landscape is to swarm with men, to be transformed from an elemental infinity to the plenitude of the registry office. This canal, this mill, these trees, these birds (Essaias van de Velde's) are linked by a crowded ferry; the overloaded boat connects the two shores and thus closes the movement of trees and water by the intention of a human movement, reducing these forces of Nature to the rank of objects and transforming the Creation into a facility.[16]

John Stuart Mill evoked a similar end more than 100 years ago, and found little satisfaction in contemplating a world with nothing left to the spontaneous activity of nature:

> ... with every rood of land brought into cultivation, which is capable of growing food for human beings; every flowery waste or natural pasture ploughed up, all quadrupeds or birds which are not domesticated for man's use exterminated as his rivals for food, every hedgerow or superfluous tree rooted out, and scarcely a place left where a wild shrub or flower could grow without being eradicated as a weed in the name of improved agriculture.[17]

Is there not a more *sustainable* concept of *sustainability*? The alternative path puts much greater emphasis on human development;

the appropriation of matter is not the measure of human progress, although a minimum economic base is necessary to free humans from suffering and fear for survival. Mill's vision sheds light on this path. He found progress or human improvement taking place where the growth of capital and population was held stationary. According to Mill, under these conditions, '[t]here would be as much scope as ever for all kinds of mental culture, and moral and social progress; as much room for improving the Art of Living and much more likelihood of its being improved, when minds ceased to be engrossed by the art of getting on'.[18] By *getting on*, Mill meant the struggle for material enrichment, while the *Art of Living* encompassed the cultivation of the graces of life, stimulation of the intellect, artistic creativity, meditative thought and character building. He included the pursuit of industrial art so long as it was showing a greater purpose than material enrichment. Mill found this path necessary to sustain balance between human actuality and naturalness of the environment.

Similar ideas have been expressed by John Kenneth Galbraith, who holds that while every country needs an economic base, having achieved it, the people should get on with *living*. In other words, there is a point on the economic growth slope where the marginal utility of an extra unit becomes negative. This point is where sustainability is threatened, and the opportunity cost is the environment and the adventure of living.

The distinctive feature of the alternative path is that material affluence is secondary or only a means in the formula. Likewise, human dignity is not dependent on government or economic dispensation, but is inherent in human nature according to the classical meaning of this term. Dignity refers to bearing that is indicative of self-respect and nobility. It equips humanity to deal with unpleasant or difficult circumstances with grace and equanimity. Consistently, the alternative path addresses the principle development issues of poverty, alienation and unemployment as deficiencies in the art of living.

Poverty is as much a question of the spirit as it is of physical circumstances. Those who are wealthy in material terms can be poor in terms of quality of life, while the materially deprived can lead rich, satisfying lives. Poverty of the soul may be most pronounced at either extreme of material conditions. It is reflected in the expressionless faces of humanity wallowing in the misery and filth of squalid urban conglomerations and in the utter hopelessness of refugees fleeing desertification and drought, but it also grays the complexions of thousands of overfed (but in fact undernourished) inhabitants of materially affluent civilizations, the victims of conformity, acquisitiveness, the drone of the media, and, most seriously, boredom.

Alienation expresses itself in violence and myriad other cruelties. It is apparent in an increasing sense of isolation, futility and insecurity. Harmony in society depends on an atmosphere of basic trust, hope and optimism; family harmony is critical to the development of these emotions and feelings. The foundations of social equipoise are essential to sustainability. Ancient Chinese tradition understood this, and Confucius articulated a system and rules to achieve this end. Likewise, child psychologists have explained the critical role of family harmony and parental security in the epigenetic sequence of human development. The effective nurturing of an infant depends on the emotional and material security of the parents. Infants require regular need fulfillment to develop basic trust in their world. This key to a healthy personality and normal socialization is readily weakened by maternal deprivation. In adults the impairment of basic trust is expressed in alienation from society and the environment.[19]

Thus, the real alternative to the dominant modernization theory implies change in values and perspectives. It requires a revolution in the common answer to the question: 'What is a good life and a good society?' Gandhi implemented a rejection of production and consumption patterns based on labour-saving technologies and the satisfaction of wants. Partly under the influence of Gandhi, and chiefly because of his own observations on the western model of unlimited quantitative growth, Fritz Schumacher developed an alternative. In *Small is Beautiful*, the chapter on 'Buddhist Economics' indicates that intellectual and moral revolution is necessary to establish another model of economic development.[20] First and most fundamental is the need to change the criterion of success for individuals and society at large. For a 'Buddhist economist' – that is, for Schumacher himself, building on the requirements of 'the Buddha's Noble Eightfold Path' – the criterion of success is the 'maximum of well-being with the minimum of consumption'. The essence of civilization is 'not in a multiplication of wants but in the purification of human character'; and, since the consumption of goods is a means to an end, 'Buddhist economics is the systematic study of how to attain given ends with the minimum means'.[21]

Schumacher gives particular attention to work and employment, and suggests giving work three basic functions, as the Buddhist philosophy does: 'to give a man a chance to utilize and develop his faculties; to enable him to overcome his ego-centeredness by joining with other people in a common task; and to bring forth the goods and services needed for a becoming existence'. Then, continues Schumacher, 'the very start of Buddhist economic planning would be planning for full employment and the primary purpose of this would in fact be employment for everyone who needs an "outside" job'.[22] In all other areas of 'development' and 'progress,' be it trade, urbanization, science and technology, and, obviously, population and the environment, the consequences for thinking and action of putting things right – that is, considering people more important than goods – would be enormous. Although Gandhi, Schumacher and other revolutionary thinkers have had some influence on 'modern economists' and decision-makers, almost everything remains to be done, intellectually and politically, to give practical meaning to this alternative vision of development, based on simplicity, moderation, wisdom and non-violence in all of its forms.

Some economists and thinkers, notably Galbraith, are exploring the conditions for the emergence of a 'good society,' according to a 'humane agenda'. Galbraith notes that everywhere 'the fortunate are now socially and politically dominant' and that 'socially desirable change is regularly denied out of well recognized self-interest'. Like Schumacher, he attributes a central and decisive role to education, and believes that 'the decisive step toward a good society is to make democracy genuine, [and] inclusive'.[23] The social dimension of the population–environment equation will thus contribute to human flourishing only if the dominant model of development is modified, and in some respects radically transformed, by another conception of what constitutes a good life and a good society.

Such notions may seem utopian in today's society, but sustainability is not a *business-as-usual* concept. It demands serious reconsideration of values, shifts in lifestyles, and a broader notion of a good society. In a democratic system such changes cannot be imposed, but must arise from an expression of the majority will to implement them. Since the majority reflects different sums of individual thinking, balance must first be sought on the personal level.

THE PERSONAL DIMENSION

Equilibrium in the personal dimension requires balance between the spiritual and the earthly experiences. It depends on a sapient knowledge of nature, and on the control of the human spirit over material perception. This sense extends beyond idealism and romanticism, which can be misused for evil purposes or to further human ambition. It demands selflessness, humility and love. An equilibrated mind operates in the material world according to the dictates of love, and instructed by nature, which, through faith, is seen to reflect the great truths of the infinite universe.

Great homage is owed to modern science and to the generalization of democracy that might one day goad respect for human rights and individual freedoms to their zenith. On the dark side, however, the obsession with material and anthropocentric processes that has ensued seems to have obscured an essential link between human life and the essence of the universe. It is the loss of life's transcendent meaning that has weighed in favour of materialism and anthropocentrism, and is the major obstacle to sustainability.

Anthropocentrism's confidence derives from humankind's discovery that it controls its own environment through increasing application of the laws of natural science. In this scientific and secular age, the vast majority of people in democratic countries share the philosophy of Thomas Hobbes for whom reality was matter in motion. Hobbes thought that the human imagination could have no scientific cognizance of the infinite nor of the immaterial.[24] But, the absence of metaphysical certainties may have also lowered the horizons of imagination and drained a sense of a larger purpose in life from daily existence.

Though an atheist, Karl Marx observed that obsessive materialism would destabilize human existence. Through an ironic interplay of ideas, he envisaged materialism and anthropocentrism operating at cross purposes, and thereby limiting the horizons of human experience. In a speech delivered in 1856, Marx implied that material progress, conceived as an end in itself, in combination with a humankind that esteems itself the master of nature, would lead the human spirit into certain vacuity, while the machine would become the controlling intelligence:

> In our days, everything seems pregnant with its contrary. Machinery gifted with the wonderful power of shortening and fructifying

human labor, we behold starving and overworking it. The new-fangled sources of wealth, by some strange weird spell, are turned into sources of want. The victories of art seem bought by loss of character . . . All our invention and progress are seen to result in endowing material forces with intellectual life, and in stultifying human life into a material force.[25]

In short, the ultimate logic of anthropocentrism and materialism is that Pinocchio becomes the master of Geppetto. Imamichi has explored this question; his theory assumes that human intention, theoretical or practical, is structured as a logical syllogism. Its classic form was elaborated in Aristotle's *Nicomachean Ethics*: the major premise is the human aim or ideal, while the minor premise is the range of free choice of means to attain the aim. Following this structure, human goal-orientation has spurred technological progress to a position of primacy over all other human aims. Thus, today, while the minor premise remains the horizon of freedom of choice, it is no longer *the means to an end*, but the goal to be realized. Because of the inversion of means and ends in human thought, goals can no longer be transcendent ideals – objectives the human spirit naturally desires – but only the natural outcomes of technological power. Material means thus control the aims of humankind, which have become the machinations of technology.[26] When humankind esteems itself capable of satisfying its aspirations, it is problematic to find a credible way to counter the increasingly monopolistic material vision.

Yet several routes remain for realizing the spontaneous and natural ideas of the human spirit. This spirit has vast resources of ideas and inspiration to deal creatively with the threatened sustainability of human civilization. Art, literature, religion, philosophy and science can offer glimpses of the transcendent spirit inspiring balance, and overruling socially negative material testimony in the human mind.

The *path of art* takes an individual to transcendence through the aesthetic experience, which is valid for both the creator and appreciator of works of art. This view has been elaborated by philosopher Noriko Hashimoto. The aesthetic experience purifies the thought and instills a sense of morality that modifies relationships to things and to other persons. The path of art cannot be objictified as a methodology; rather, it is a path unto itself, to be experienced personally. According to Hashimoto, a fundamental philosophical belief in the Orient is that human beings are a part of nature, linked by spirit to the vital power of the universe. Individuals bring this spirit into their experience through efforts to understand nature's essence in material things, be it a blade of grass, or a stone beside the road. In the eighth century, the Chinese painter Wan Wei said that '[w]hen a painter draws a mountain, he must grasp and draw the "spiritus" of the mountain, not identify which mountain it is'. Going to the mountain and resting there encourages this understanding. Similarly, the 'path of flower arrangement' is to gather the flowers and, with infinite solicitude and an eye to artistic composition, to place them in vessels of water to expose their temporally limited beauty. Through flower composition, one can learn the significance of spacial and temporal relations in nature, as well as in relations between people. Thus, through creation and appreciation of art, infused with the *spiritus* of nature, one finds unity between material life, ethics and religion.[27]

To a transcendent vision, fantasy and muse are the handmaidens of reason. In fairy tales is to be divined the wonder of things such as stones, wood, trees, grass, rivers, bread and wine. Imagination may capture sunlight turning the dew on leaves into diamonds, and creating out of the common gray spider web a fabric of exquisite and glimmering design. With such appreciative insights the heart may approach nature, pursue adventure, and find ultimate good.

In the medium of the fairy story, J.R.R. Tolkien found passage to a different dimension of truth, and therein a promising route to an enchanting perspective on the environment. By definition, true fantasy ends in joy and has the *inner consistency of reality*. For lack of a better term, Tolkien offers the word *eucatastrophe* to contrast his concept of the fairy story with the conception of drama as a form of *tragedy* ending in sadness. While not denying the existence of sorrow and failure, in the end the fairy story offers *evangelium*, which Tolkien defines as sudden and miraculous grace. If the tale has the *inner consistency of reality*, it will somehow reveal an abiding transcendent truth that will uplift the horizons of human hope and feed a sense of optimism. The fairy tale is not only truth in the secondary world, but as an eucatastrophe, it offers a glimpse of a far off gleam or echo of divine grace in the real world.[28]

Notions of dignity and nobility may be cultivated in literature and art. Many works offer examples of love, nobility and responsibility reflected in human lives and projected on other forms of life, awakening empathy and respect. Through imagery, the writer also conveys the process by which such fine qualities inherent in humanity can be obscured. The chasm that has developed between man and nature, and the decline in human dignity that occurs in an excessively materialistic society, have been evoked by D.H. Lawrence, in his reflections of a horsewoman studying her stallion:

> But, now, where is the flame of dangerous, forward pressing nobility in man? Dead, dead, guttering out in a stink of self sacrifice whose feeble light is a light of exhaustion and laissez-faire. And the horse, is he to go on carrying man forward into this? – this gutter? No! Man wisely invents motor cars and other machines, automobile and locomotive. The horse is superannuated, for man . . . she knew that the horse, born to serve nobly, [had waited] in vain for someone noble to serve. His spirit knew that nobility had gone out of men. And this left him high and dry, in a sort of despair . . .[29]

Religion, in many different ways, offers counterpoints to the material perspective. As de Tocqueville pointed out,

> . . . there no religion that does not place the man's desires above and beyond the treasurers of earth and that does not naturally raise his soul to regions far above those of that senses. Nor is there any which does not impose on man some duties towards his kind and thus draw him at times from the contemplation of himself.[30]

There are many examples of how religious thought can contribute to building empathy and an environmental ethic through inspired reason and imagery.[31]

Buddhism emphasizes the purification and development of the mind. It reveals different *paths* for seeking an understanding of reality, of the fundamental law of nature, and ultimately, of supreme liberation. The *natural path* reveals the material world to be empty of intrinsic and independent essence. Primordially, reality is based on the principle of the interdependence of *things* with thought, and with conceptual designations. This interdependence reveals a balance between the inner and outer worlds. In its purest state, the inner self is imbued with genuine compassion, and with a strong sense of responsibility for the well-being of all life. This is the source of balance and hope.[32]

Islamic thought offers a metaphysic of nature and love. The soul that is sustained by the Koran regards nature as an integral part of its religious universe, sharing in its earthly life, and, in a sense, even in its ultimate destiny: 'nature as being ultimately a theophany which veils and reveals God'.[33] The traditional Islamic view of the natural environment is based on the assumption of an inherent relationship between the material manifestations of life, and the Divine Environment which sustains them. God is the ultimate environment (*Muhit*) that surrounds and encompasses humanity. The enemies of this environment are modern science and predatory economic humanism, which limit knowledge to the boundaries of empiricism and rationalism, thereby demystifying human nature. No longer an enchanted world, nature becomes a collection of objects to exploit and to consume. It is then necessary to rediscover 'the nexus between the Spirit and nature and awareness of the sacred quality of the works of the Supreme Artisan'.[34]

Asceticism blended with a love of learning is the Judaic prescription for gaining a state of mind that favours inner balance and defends the individual against the grinding materialism of outward concerns. In accordance with scripture, the infinite struggle of humankind in nature is the struggle between the inner and outer person for the triumph of spirituality over the senses. Survival in the here and now depends on a regime of plain living. That simple life is to be filled with study of scriptures and other holy works, the distillates of revelatory experience. This is the way to cultivate the inner life that will counter the strains and seductions of prosperity or oppression. Study hollows out a pocket of tranquillity in which to take refuge.[35]

The Christian message is decidedly anti-materialist. This message is conveyed in a number of different ways by Christ in his parables and the Sermon on the Mount, as well as in other places in the gospels. Materialism is corrupting (Matthew 7); it blocks the consciousness of heaven or inner serenity (Mark 10:21–25); and it is insufficient to sustain a full life experience (Luke 4:2–4).[36] Nature provides the examples that Christ uses to direct human attention away from its fascination with creature comforts, adornments and acquisitiveness. In the Sermon on the Mount, he says: 'Take no thought for your life, what ye shall eat, or what ye shall drink; nor yet for your body what ye shall put on . . . Which of you by taking thought can add one cubit unto his stature' (Matthew 6:26–27). Nature is the sublime example that guides humankind to the good life and to an appreciation of divine sustenance and supply. The birds demonstrate nature's provision of food, and the lilies reflect beauty emerging from the soul, and elsewhere throughout the Bible the trees, the grass, the vine, the lion, the lamb and the wolf, the rocks and still waters, and the sand all have symbolic import for human life.

Spirituality is inherent in traditional cultures of the native American Indians as well. Although the different tribes had many different ways of reflecting this, for all of them spirituality was manifest in their social organizations, world views, rituals, lifestyles and modes of subsistence. Sensitivity and intimacy with the inanimate and animate populations of the environment were intrinsic to their life experience. These relations nourished love, respect, thanksgiving and responsibility as cornerstones of environmental ethics.[37] The native American vision sees in nature manifestations of the Great Spirit, and a lake is the smile of the Great Spirit. The evergreen tree is a timeless message of cosmic harmony; as the great tree of peace, it signifies the Great Law of Equity and Justice. Closely associated with this law is the profound sense that humanity is part of the great family.[38] More than a half century ago, Luther Standing Bear wrote:

> We are of the soil and the soil is of us. We love birds and beasts that grew with us on this soil. They drank the same water and breathed the same air. We are all one in nature. Believing so, there was in our hearts a great peace and willing kindness for all living growing things.[39]

From the Confucian perspective, the full realization of human potential is anthropocosmic. Nature provides the material sustenance for life and the inspiration for sustainability. The diurnal rotations of the earth and the seasonal transformations are enduring patterns of transformation, regularity, balance and harmony. The sense of nature as a home empowers one to find ultimate meaning in ordinary life, to cultivate a regularized, balanced and harmonious lifestyle, and to regard the secular as sacred. In the order of things, one achieves such a lifestyle by cultivating loving and respectful relationships in family, community and state.[40] The dichotomies between matter/spirit, man/nature, and creativity/creature must be transcended to allow supreme values such as the sanctity of the earth, the continuity of being, and the beneficent interactions between humanity and nature, and between humankind and heaven, to receive the salience they deserve in philosophy, theology and life itself.[41]

Nurturing the human spirit and finding the roots of a sustainable balance depend on humankind's capacity to find the time and space for meditation in solitude amidst the beauty of nature. In Mill's view, the biggest threat posed for society by anthropocentrism and imbalance was the loss of time and space for this activity:

> A population may be too crowded, though all be amply supplied with food and raiment. It is not good for man to be kept perforce at all times in the presence of his species. A world from which solitude is extirpated, is a very poor ideal. Solitude, in the sense of being often alone, is essential to any depth of meditation or of character; and solitude in the presence of natural beauty and grandeur, is the cradle of thoughts and aspirations which are not only good for the individual, but which society could ill do without.[42]

On the physical plane, science has catapulted the individual into a modern, global, technological civilization under a regime of enlightened materiality.

Yet with the assistance of faith and imagination, science can also lead humanity to equilibrium. Quantum physics and higher mathematics, for example, allow one to take a mental journey on an ultra-microscopic trajectory through the universe of the atom, through that of the quark, and onward in an odyssey to smaller and smaller particles, until one discovers that matter, and then energy, seem to disappear. Finally, one stands on the frontier of science and religion and glimpses the eternal cosmos, perhaps as an infinite 'light' illuminating the human intellect through revelation and inspiration. Physicists reveal this universe as one of cosmic order and infinite freedom. Human thoughts, upon returning to these roots, find harmony, nobility and purpose. Einstein defined such experience as cosmic religious feeling which

> ... takes the form of a rapturous amazement at the harmony of natural law, which reveals an intelligence of such superiority that, compared with it, all the systematic thinking and acting of human beings is an utterly insignificant reflection. This feeling is the guiding principle of ... life and work, in so far as [individuals] succeed in keeping [themselves] from the shackles of selfish desire. It is beyond question closely akin to that which has possessed the religious geniuses of all ages.[43]

The sublimity and order that reveal themselves both in the spirit of nature and in the world of thought offer infinite strength, possibilities and purpose to humanity as it experiences the futility of material aims, desires or necessities. Einstein offered that it is the high destiny of individuals to serve rather than to rule or to impose themselves in any other way.

Translating from this cosmic point of view to a more earthly vision, the environment can be imagined as a projection of human thought whose repair begins in the human mind. Detoxification, waste removal, recycling, conservation, and sexually responsible behavioural most effectively begin in minds imbued with strong affection for humanity and an appreciation for life that reflects the beauty and grandeur of nature and the universe. Social pollutants can be neutralized by humility, a noble life purpose, and a lifestyle guided by *receptivity* to cosmic or spiritual feelings.

CONCLUSION

Each of the three dimensions is relevant to achieving a sustainable population–environment equilibrium. The personal plane, however, is primordial for effective political policy. The legitimacy of democratic government is based on the principles that sovereignty rests with the people, that the state serves the needs and interests of the individual, and that political action reflects the majority will. The grounds for political obligation are consent, moral right and justice. The substance of law, therefore, reflects what people value. Genuine politics, according to Vaclav Havel, consists in concern for service to others and the world. Its roots are noble, because it is a 'higher' responsibility to the whole that emerges

> ... out of a conscious or subconscious certainty that our death ends nothing, because everything is forever being recorded and evaluated somewhere else, somewhere 'above us' in ... the 'memory of Being' an integral aspect of the secret order of the cosmos, of nature, of life, which believers call God ...[44]

Equilibrium thus begins with the individual's compelling concern for and appreciation of his or her environment, whether in the realm of thoughts, the social community or the physical surroundings. With the globalization of democratic capitalism predicated on individualism and human rights and the sanctities of private property and the market, it is more essential than ever that individual thought be steeped in reverence for life and nature.

Whether, as Emerson suggested, 'things are in the saddle and ride man', or whether man and nature are in the saddle, is humankind's decision. It is not surprising, for example, that a country with a capital known as the garden city of the world includes among its greatest heroes an author of fairy tales and a theologian/philosopher. In the same country, which has a high income per capita, markets close daily at 5 p.m. (2 p.m. on Saturday for the weekend), and offices at 4 p.m. In this country, bicyclists rule the roads, and there are neither homeless people nor socially justifiable reasons for misery. It is not surprising either that in another country, which is pioneering the international cause for protecting the environment, every student must be trained in ethics, civility or religion, and the national anthem is a hymn to the grandeur of mountains rising from the sea.

The cynic or the skeptic will find many reasons to explain away such apparent aberrations in the modern world, including the homogeneity and small size of these countries' populations. The point, however, is not the physical circumstances, which do not differ dramatically in many parts of the world. The point is that the apparent harmony in some countries or regions of the contemporary world manifests the predominance of some degree of faith in a higher intelligence and/or social responsibility. These societies are disciplined by an ethos that tempers *laissez-faire*, and philosophies that regard life as something larger than the marketplace and its opportunities. These conditions can exist everywhere, independent of material affluence, if we so choose.

Notes

1. Lynton Keith Caldwell, *International Environmental Policy: Emergence and Dimensions*, 2nd edn (Durham and London: Duke University Press, 1990), 8.
2. Albert Einstein, 'Physics and Reality', *Ideas and Opinions*, based on *Mein Weltbild*, edited by Carl Seelig, and other sources (New York, Bonanza Books, 1984), 290–1. See also remarks on *Bertrand Russell's Theory of Knowledge*, ibid., 22–3.
3. Interview with author, Tyngsboro, Massachusetts, March 1994.
4. Richard B. Mancke, *The Failure of U.S. Energy Policy* (New York: Columbia University Press, 1974), 35–45.
5. See Robert Clark, *Power and Policy in the Third World*, 4th edn (New York: Macmillan, 1991); Robert E. Gamer, *The Developing Nations:*

   A Comparative Perspective, 2nd edn (New York: W. C. Brown, 1988);
   and Robert Nisbet, *History of the Idea of Progress* (New York: Basic
   Books, 1980).

6. Adam Smith, *The Wealth of Nations* (New York: Modern Library, 1994),
   14.

7. *Ibid.*, 373.

8. *Ibid.*, 485.

9. John Locke, *Two Treatises of Government*, Cambridge texts in history
   of political thought, ed. Peter Laslett (New York: Cambridge University Press, 1994), 323–4.

10. *Ibid.*, 106, 357.

11. *Op. cit.*, Smith, 839–40.

12. Tomonobu Imamichi, 'The Concept of an Eco-Ethics and the Devel-
    opment of Moral Thoughts', *Man and Nature*, ed. George McLeon
    (New York: University Press of America, 1984), 213–14.

13. Karl Marx, *Capital*, Vol. 1 (London: Penguin Classics, 1990), 163–77.

14. *Ibid.*, 141.

15. Claude Ake, 'Market and Social Darwinism', in *Ethical and Spiritual
    Dimensions of Social Progress* (New York: United Nations, 1995), 43.

16. Roland Barthes, *A Barthes Reader* (New York: Hill & Wang, 1995),
    62–3.

17. John Stuart Mill, *Principles of Political Economy* (New York: Oxford
    University Press, 1994), 129.

18. *Ibid.*

19. Erik H. Erikson, *Identity and the Life Cycle: Selected Papers, 1*, vol. 1,
    no. 1 (International University Press, 1959), 52–7, and commentary
    by Dr Peter Kukla, former Associate Clinical Professor of Pathology
    and Clinical Fellow in Psychiatry at Harvard University.

20. E.F. Schumacher, *Small is Beautiful: Economics as if People Mattered*
    (London: Harper Perennial, 1975).

21. *Ibid.*, 59 and 67.

22. *Ibid*, 58 and 60.

23. John Kenneth Galbraith, *The Good Society* (London: Sinclair-Stevenson,
    1996), 2, 4 and 139.

24. Thomas Hobbes, *Leviathan* [1651], Part I, Ch. 3 (New York: Penguin,
    1985), 94–5.

25. Karl Marx and Friedrich Engels, *Selected Works*, Vol. 3 (Moscow:
    Progress Publishers, 1969), 500; as cited in John Young, *Sustaining
    the Earth* (Cambridge: Harvard University Press, 1990), 66.

26. Imamichi, 212–13.

27. Noriko Hashimoto, *Aesthetic Reflection on the Path of Art in Japan*,
    paper presented at Haus der Kuluteren der Welt, 9–10 October 1996.

28. J.R.R. Tolkien, 'On Fairy-Stories', in *The Tolkien Reader* (New York:
    Ballantine, 1966), 85–6.

29. D.H. Lawrence, *St. Mawr & The Man Who Died* (New York: Vintage,
    1953), 68.

30. de Tocqueville, Vol. II, 22.

31. A symposium held in at Middlebury College brought together leading
    theologians or exponents of several of the worlds major religions to

discuss the relevance of religion and spirit to the environment. See
Steven C. Rockefeller and John C. Elder, *Spirit and Nature: Why the
Environment is a Religious Issue* (Boston: Beacon Press, 1992).

32. Tenzin Gzatso, His Holiness the 14th Dalai Lama, 'A Tibetan Buddhist
Perspective on Spirit and Nature', in *Spirit and Nature*, 115.
33. Seyyed Hossein Nasr, 'Islam and the Environment Crisis', in *Spirit
and Nature*, 89.
34. *Ibid.*, 105–6.
35. Ismar Schrosch, 'Learning to Live with Less: A Jewish Perspective',
in *Spirit and Nature*, 36.
36. Derived from St James translation of the Bible.
37. John A. Grim, 'Native North American Worldviews and Ecology', in
Worldviews and Ecology (Maryknoll, New York: Orbis Books, 1994),
41–53.
38. Audrey Schenadoah, 'A Tradition of Thanksgiving', in *Spirit and Nature*,
17.
39. Luther Standing Bear, *Land of the Spotted Eagle* (Lincoln: University
of Nevada Press, 1933), 45.
40. Tu Wei-ming, 'Confucianism', in *Our Religions* (San Francisco:
HarperCollins, 1993), 144–5.
41. Tu Wei-Ming, 'Beyond the Enlightenment Mentality', in *Worldviews
and Ecology.*
42. *Op. cit.*, Mill, 128–9.
43. Einstein, *op. cit.*, 40.
44. Vaclav Havel, 'Paradise Lost', *The New York Review of Books* (8 April
1992), 6.

Part I
Factors in the Equation

2 Population and Environment Relationships in Developing Countries: Recent Approaches and Methods

Catherine M. Marquette and
Richard E. Bilsborrow

A DIVERSITY OF PERSPECTIVES

Research and interest in the links between population dynamics and environmental change were given renewed impetus by the United Nations Conference on Environment and Development, held in Rio de Janeiro in 1992. The conference summary statement, Agenda 21, recommended the development and dissemination of knowledge on the links between demographic trends and sustainable development, including environmental impacts.[1] Despite achieving this final consensus, the discussion about linkages between population and environment was highly charged. Grassroots development, environmental, and women's groups strove to keep the wider social and economic contexts in which population and environment relationships occur in the forefront, contrary to the more focused interests of many population groups.

This diversity of opinion and approaches generally characterizes discussion of population and environment relationships in both public and academic contexts. The following review tries to capture some of this diversity by briefly considering some of the different perspectives on the topic (see Figure 2.1). Although presented separately, many of these perspectives overlap, and many studies reflect the influence of more than one perspective. This review merely

Figure 2.1   Some conceptual approaches to population and environment relationships

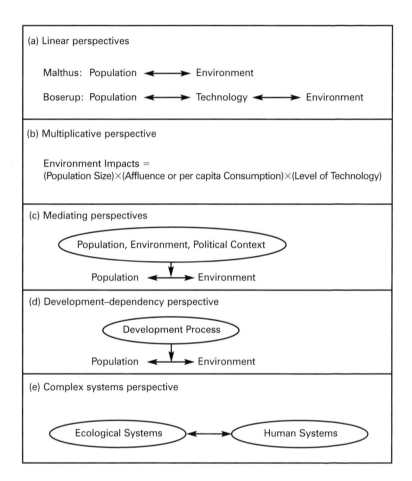

begins to unravel some of the various strands that have shaped thinking on the topic, both historically and in the present. The objective is to stimulate further thinking about and analysis of conceptual approaches, and to encourage their more explicit formulation in future research.

Linear Views: Malthus and Boserup

Neither Malthus nor Boserup specifically address population–environment relations; rather, they limit their focus to the narrow topics of land-use and food production. Their ideas do, however, effectively represent the two dominant historical viewpoints on these topics, and implications for the general linkages between population and resources are frequently inferred from their work. Both of their perspectives emphasize the reciprocal, linear and direct relationships that exist between population and the environment.

Malthusian theory[2] stresses that the growth of human populations always tends to outstrip the productive capabilities of land resources. The result is that 'positive' checks, such as famine and increased mortality, or preventative checks, such as postponement of marriage and limitation of family size, work to reduce population growth. Malthus suggests that population demands therefore directly limit the availability of resources and that resources, in turn, directly restrict population growth. Malthusian theory, formulated before the agricultural revolution, presumes that the productivity of environmental resources such as land is fixed.

Malthus did not foresee the important technological advances that have accompanied modernization. Writing after the agricultural and industrial revolutions, Boserup does take this technological change into account. She suggests that population growth and the increased population density that results 'induce' technological changes, such as the use of plows or fertilizer, which allow food production to keep pace with population growth. Again, reciprocal linear relationships between population, technological change in agriculture, and environmental change are suggested.[3]

Malthusian ideas have informed much of the subsequent discourse on population–environment relations, including numerous descriptive studies on demographic and ecological trends.[4] The Malthusian viewpoint has also had an influence on the development of the concept of 'carrying capacity', which has lead to several global and national projection and modelling exercises.[5] The Boserupian perspective has likewise influenced global and regional research that examines the relationship between population growth and changes in agricultural production.[6]

Multiplicative Perspectives: The 'IPAT' Equation

Another current line of thought sees population size as interacting in a multiplicative way with other factors to generate impacts on the environment. One of the most frequently used multiplier approaches is the so-called 'IPAT' equation in which:

> Environmental impacts = (Population) × (Level of affluence or per capita consumption) × (Level of technology)

or

$$I = P \times A \times T^7$$

The IPAT equation treats the combined interactions among population, consumption and technology as important determinants of environmental change, rather than addressing only their independent effects.

Shaw has proposed an alternative multiplicative scheme in which the interactive effects between population, consumption and technology are specified in more detail. He distinguishes between ultimate causes, or the driving forces behind environmental impacts, and aggravating factors. In the case of environmental degradation, consumption and technology are the ultimate causes, while population is an aggravating factor that increases the intensity of the impacts of the ultimate causes.[8]

Mediating Perspectives

Numerous studies focus on the context in which population and environment relationships occur, or the social, cultural, institutional and political factors that mediate relationships. Since a wide range of mediating factors may be considered, the various studies that have been carried out under this approach are also diverse. Bilsborrow has elaborated a mediating framework for understanding the impacts of population growth on land use and agricultural production in rural areas of Latin America. This framework considers how socioeconomic conditions such as poverty, government policies and market demands determine whether population growth leads to technological change in agriculture, soil degradation or out-migration.[9] Other mediating viewpoints focus more exclusively on social and cultural factors, rather than economic and policy ones, as mediators of population and environment relations.[10] In contrast to the direct relationship that exists between other animals and the environment, these viewpoints emphasize that social organization and culture filter and focus the relationship between human populations and their environment. Environmental change is thus viewed as a social as well as a natural process.[11]

Development–Dependency Perspectives

Another perspective collapses all social, cultural and institutional factors that mediate population–environment relationships into the larger concept of 'development', and then focuses on the way in which development processes mediate population–environment relations. Particular emphasis is placed on development trends that have kept the South 'dependent' on the North, such as mercantile exploitation and the export of natural resources to manufacturing centres in the North. This 'dependency perspective' stresses the overwhelming role that common international political and economic forces play in shaping both demographic factors such as population growth, and environmental outcomes such as degradation, in developing countries.[12] This approach further suggests that even major global environmental problems (ozone depletion, global warming, toxic wastes and biodiversity loss) are the direct result of the prevailing model of development, which is based on rapid industrialization.[13] The spread of this industrial-led pattern of development across low- and middle-income countries is viewed as compounding negative global environmental trends.

Complex System Perspectives

Another approach considers both mediating factors and environment and population in a structured way, as a complex of inter-related systems. This approach aims to understand how ecological and human-driven systems (sociocultural, demographic and economic systems) interconnect to form larger 'socio-ecological systems'[14] within which population and environment relationships are embedded. Population ecology, human ecology and cultural ecology – all sub-disciplines within anthropology – adopt this approach in studying how human systems reflect adaptations to a given ecosystem, as well as how human systems may shape natural ecosystems.[15] This approach also accounts for large-scale structural changes, such as developmental processes, which may cause radical shifts in existing human and ecological systems and the relationships between them.[16]

Strengths and Weaknesses of the Contending Perspectives

Each of the perspectives discussed above has strengths and weaknesses with respect to the conceptual relationships and methodological approaches implied. Malthusian and Boserupian approaches provide the most straightforward theory on population–environment relations, presenting clear propositions about the nature of these relationships. However, their contrasting conclusions have frequently turned research on population and environment into a battleground for an ideological war waged between the so-called 'neomalthusians' and the 'cornucopians'.[17] In addition, it is difficult to operationalize both Malthusian and Boserupian concepts (for example population pressure or technological change) as variables that can actually be measured and studied.

Multiplicative approaches such as the IPAT equation, in contrast, provide a calculable formula for estimating environmental impacts. On the other hand, the IPAT equation may reduce complex phenomena to quantifiable generalities (broad measures of population, consumption and technology) thereby missing the local-level characteristics of resource use that may be key to understanding population and environment linkages. Mediating approaches are more sensitive to local and contextual factors that may shape population and environment linkages. Yet the idea of 'mediation' is ambiguous, since the direction, priority and nature of the interactions between 'mediating' socioeconomic factors and population and environment relationships is not always clear. Complex systems provide further specification of these mediations, but demand comprehensive information from different sectors and at different levels of aggregation, and this may be difficult to obtain, much less process and analyze.

In all of these studies, the concept of 'population' has been limited largely to a focus on population growth.[18] The mediating perspectives do, however, consider other dimensions of population in relation to environmental change, including migration and the spatial distribution of population, land-tenure patterns, household formation and household-level demographic characteristics (size and structure), and the reciprocal impacts of environmental degradation on population health. The concept of 'environment', in contrast, has been defined in more diverse ways. The environmental variables most frequently considered centre on aspects of specific resources (water and air quality, deforestation, land degradation). The precise variables used range from specific, quantitative measures of pollution, soil loss or deforestation, to more qualitative and impressionistic estimations of overall 'deterioration'. The current concepts upon which environmental variables are based generally adhere to a capitalistic model, which views the environment as having value mainly in relation to human use (use-value).[19] Alternative notions based on other paradigms – for example, views that recognize the environment as having value that transcends human use, such as for the maintenance of other species – have not been widely integrated into current studies.

With the exception of more anthropologically-oriented studies, investigators generally define *a priori* the concepts of population and environment, and the relationships between them. The perceptions of affected populations with respect to the boundaries of their environment, the impact of their activities on the environment, and the reciprocal impacts of environmental change on their own lives are generally not taken into account.[20] This practice continues despite the fact that environmental perceptions may be a key factor linking populations to environmental change.

DIVERSITY IN LEVEL OF ANALYSIS

In addition to their differing conceptual approaches, studies of population–environment relationships vary according to the geographic level of analysis. Macro-level studies involve large units of analysis such as the entire globe, developing regions, countries, or regions within countries. Micro-level analysis, in contrast, involves smaller units such as households, families or specific communities. Macro- and micro-level studies therefore differ with respect to their data needs, the methodological approaches utilized, the possibilities they offer for making generalizations and drawing conclusions, and the information they can provide for policy formulation.

Macro-level research generally draws on existing aggregate data, involves quantitative approaches that make global, cross-regional or cross-country assessments, and produces conclusions that provide information on general relationships that apply to large populations or geographic regions. Data from global studies is thus useful for elaborating international and national policies. Micro-level research, in contrast, requires disaggregated data, frequently involves qualitative methods and specialized data collection, and produces less generalizable conclusions that relate to small, specific populations or communities. Micro-level research can, however, draw upon more detailed information to identify how social, economic, cultural and institutional factors influence the nature of population–environment relationships in different contexts.

Although discussed separately, macro- and micro-level studies may be effectively combined to give a more comprehensive understanding of linkages. Macro-level studies may identify broad hypotheses that can then be tested at the micro-level. For example, linkages between global consumption patterns and climate change might be explored at both national and sub-national levels to identify different patterns between and within countries. The majority of recent research on population and the environment has, however, been carried out at the macro level. Ehrlich's examination of the 'population bomb' and the 'population explosion',[21] as well as the study of global 'limits to growth'[22] have attracted popular as well as academic attention for the last three decades. These global studies by natural scientists are perhaps the best known research to date on population and environment relationships. Taking the lead from their natural science counterparts, demographers and economists have also tended to consider the macro-level impact of population growth on global food supply, climate change, or natural resource depletion.[23]

Many of these macro-level studies describe rather than explain the causal linkages between population and environmental change. Cross-sectional quantitative or qualitative data and relationships are generally presented, and cause and effect relationships over time are simply inferred. As a result, these largely descriptive macro-studies provide little insight into the causal relationships that link population dynamics and environmental outcomes at the household or community level and within critical regional ecosystems, such as tropical forests, mountain areas, dryland savannas and coastal regions.[24] Further micro-level study at the sub-national, community and household levels is needed to explore these linkages.[25]

DATA ISSUES

Existing information has only begun to be exploited in the analysis of population and environment relationships. For example, agricultural census surveys and population census data have been used for this purpose in Latin America,[26] and the potential for similar utilization of existing population and agricultural census data exists in other regions as well.[27] Several existing national and regional databases also contain both population and environmental data that might be used in future macro- and micro-level research. These include the World Bank Living Standard Measurement Survey (LSMS) data, collected in about a dozen developing countries, UNESCO Man in the Biosphere (MAB) Program data, and data collected by the Consortium for International Earth Science Information Network (CIESIN).

Geographic Information Systems (GIS) offer an important tool for combining and analyzing demographic and environmental information, and increasing efforts are being made to utilize this tool.[28] Some existing databases that already use GIS to link relevant information on population and the environment include the Global Environmental Monitoring System (GEMS) and the Global Resource Information Data Base (GRID), both created by the United Nations Environment Program (UNEP), and the Famine Early Warning System (FEWS) maintained by the United States Agency for International Development (USAID). Further scope still exists for analysis of this linked information at the global, national and subnational levels.

Local-level population–environment monitoring systems (PEMS) have also been set up in some developing countries using GIS.[29] These systems are prospectively collecting demographic, health, socioeconomic and environmental data at the local level for integration into GIS systems. Increased use of mapping and GIS technologies has also begun to occur among local community groups themselves in an effort to learn more about population and resource relationships that affect them.[30] Use of geographic positioning systems (GPS) in conjunction with GIS is allowing this local-level information to be combined with higher level maps and information.[31] A wealth of important information on population and environment relationships thus exists at the community and local levels, and this data can be utilized for local-level analyses or aggregated for use at the sub-national or national levels.

In addition to the availability of existing data, new information will inevitably be collected as well. Given this fact, it is important to recognize the need to make population and environmental data more comparable in the future.[32] For example, population data from censuses and surveys are collected on the basis of political or administrative units; it therefore may not match environmental data which are collected based on ecosystem, topographic or climatic zone boundaries. The future investigation of population–environment relationships would thus benefit substantially from the collection of demographic data in a way that would facilitate analysis on the basis of ecological or climatic zone.[33] Greater comparability between population and geographic data will also facilitate the application of GIS to analyze relationships.[34]

Ongoing national or international data-collection programmes might also be adapted to allow the integrated analysis of population, land-use, economic and environmental trends in the future.[35] For example, it has been proposed that an environmental module be included in the ongoing Demographic and Health Surveys (DHS) currently being carried out by Macro International in numerous developing countries.[36] Such an environmental module would afford the linkage of environmental information with the extensive household and community-level data already collected by the DHS.

IS A GENERAL THEORY OF POPULATION AND ENVIRONMENT RELATIONSHIPS POSSIBLE?

The majority of population and environment relationships are played out as local dramas, and must first be fully understood in this context. Even global environmental impacts (for example loss of biodiversity or global warming) have their roots in processes played out within regions, communities and households, and the localized character of other processes such as soil degradation, deforestation and urban environmental deterioration is readily apparent. Research would therefore benefit greatly from a greater emphasis on what the sociologist Robert Merton termed middle-range theory and research, which attempts to explain a limited phenomena in a specific context as well as possible.[37]

The basic ingredient for moving towards such middle-range theory and research is again micro-level research. As noted above, much of the resources and attention that have gone to the study of population and environment relationships have gone to expensive, large-scale, multidisciplinary studies or sophisticated macro-level modelling and simulation exercises headed by established academics with large research teams. The Global Environmental Fund, the Human Dimensions for Global Environmental Change Project, the World Bank, USAID, the United Nations (including UNFPA, UNDP and UNEP), and private foundations such as the MacArthur Fund have all directed many of their resources towards this type of macro-level research.

In the future, more of these funds should flow towards micro-level studies. This includes more support for studies carried out by graduate students, community groups, and local-level non-governmental organizations. Gaining greater understanding of population and environment relationships depends on the accumulation of these more humble, middle-range studies, rather than on grandiose and costly multidisciplinary studies and global projection exercises. The cost-effectiveness of this new emphasis in terms of the amount of information to be gained is obvious. This is particularly relevant in the current environment of budget cutting and restricted resources.

Ultimately, a more complete vision of population and environment relationships, including a better understanding of global relationships, may emerge from the accumulation of micro-level studies and empirical evidence as a substitute for researchers' assumptions. Most importantly, micro-level studies offer a way to accumulate and apply, little by little, information for constructing realistic policies affecting population and environment relationships at the household, community, regional and ultimately national level. International policy-making may be built upon this alternative foundation of grass-roots involvement, rather than global pronouncements of doom. The adage 'think globally, act locally' has particular significance in this context. For the near future, the 'bottom-up' approach of micro-level study, rather than the 'trickle-down' approach of macro-level study, should be the driving force in social science research on population and environment relationships.

Notes

1. United Nations, *Report of the United Nations Conference on Environment and Development (UNCED)*, Rio de Janeiro, 3–14 June 1992, Vol. 1 (New York: United Nations, 1993).
2. T. Malthus, *First Essay on Population* [1798[and *Second Essay on Population* [1803] (New York: Modern Library and Random House, 1960).
3. E. Boserup, *The Conditions of Agricultural Growth* (Chicago: Aldine, 1965); E. Boserup, 'Environment, Population and Technology in Primitive Societies', *Population and Development Review* 2 (1976): 21–36; E. Boserup, *Population and Technological Change* (Chicago: University of Chicago Press, 1981).

4. See, for example, L. Brown *et al.*, 'Twenty-Two Dimensions of the Population Problem', World Watch paper no. 5 (Washington, DC: Worldwatch Institute, 1976); P. Ehrlich, *The Population Bomb* (New York: Ballantine, 1968); P. Ehrlich, A. Ehrlich and G. Daily, 'Food Security, Population and Environment', *Population and Development Review* 19 (1993): 1–32; P. Ehrlich and A. Ehrlich, *Ecoscience: Population, Resources, Environment* (San Francisco: W.H. Freemen, 1977); P. Ehrlich and A. Ehrlich, *The Population Explosion* (New York: Simon & Schuster, 1990); P. Ehrlich and J. Holdren, 'The Impact of Population Growth', *Science* 171 (1971): 1212–7; P. Ehrlich and J. Holdren, 'Human Population and the Global Environment', *American Scientist* 62 (1974): 282–92; E. Eckholm, *Losing Ground* (New York: W.W. Norton and World Watch Institute, 1976); G. Hardin, 'The Tragedy of the Commons', *Science* 162 (1968): 1248.
5. See, for example, J.E. Cohen, *How Many People Can the Earth Support?* (New York: Norton, 1995); G. Higgins, A. Kassam, L. Naiken, G. Fischer and M. Shah, *Potential Population Supporting Capacity of Lands in the Developing World* (Rome: United Nations Food and Agriculture Organization (FAO), 1982); W. Lutz, 'Population, Environment and Development: A Case Study of Mauritius', *Options* (1991): 11–6; D. Meadows *et al.*, *The Limits to Growth* (New York: Universe Books, 1972); D. Meadows et al., *Beyond the Limits: Confronting Global Collapse, Envisioning a Sustainable Future* (Post Mills, Vermont: Chelsea Green, 1992).
6. J. Simon, *The Ultimate Resource* (Princeton: Princeton University Press, 1981); J. Simon, *Population Matters: People, Resources, Environment and Immigration* (New Brunswick: Transaction, 1990).
7. See, for example, Ehrlich and Holdren, 'Impacts'; Ehrlich and Holdren, 'Human Population'; P. Harrison, *The Third Revolution: Environment, Population, and a Sustainable World* (London and New York: I.B. Tauris, 1992); B. Commoner, 'Rapid Population Growth and Environmental Stress' in *Consequences of Rapid Population Growth in Developing Countries: Proceedings of the United Nations/Institute National d'Etudes Demographique Expert Group Meeting* (New York: United Nations, 1991), 161–90; B. Commoner, 'Population, Development and the Environment: Trends and Key Issues in the Developed Countries', paper presented at the United Nations Expert Meeting on Population, Environment and Development, New York, 20–24 January 1992.
8. P. Shaw, 'Paradox of Population Growth and Environmental Degradation', paper presented to the American Association for the Advancement of Science (AAAS) Annual Meeting, San Francisco, California, 14–19 January 1989; P. Shaw, 'Population, Environment and Women: An Analytical Framework', paper prepared for the 1989 United Nations Population Fund (UNFPA) Inter-Agency Consultative Meeting, New York, 6 March 1989; P. Shaw, 'Rapid Population Growth and Environmental Degradation: Ultimate Versus Proximate Factors', *Environmental Conservation* 16 (1989); P. Shaw, 'The Impact of Population Growth on Environment: The Debate Heats Up', *Environmental Impact Assessment Review* 12 (1992). See also D. Hogan, 'The Impact of Popu-

lation Growth on the Physical Environment', *European Journal of Population* 8 (1992): 109–23.

9. R. Bilsborrow, 'Population Growth, Internal Migration and Environmental Degradation in Rural Areas of Developing Countries', *European Journal of Population* 8 (1992): 125–48; R. Bilsborrow, 'Population, Development and Deforestation: Some Recent Evidence', paper presented at the United Nations Expert Group Meeting on Population, Environment and Development, New York, 20–24 January 1992.

10. See, for example, G. McNicoll, 'Social Organization and Ecological Stability Under Demographic Stress', Population Council working paper no. 11 (New York: The Population Council, 1990); D. Hogan, 'Impact of Population Growth'; M. Sahlins, *Stone Age Economics* (Chicago and New York: Aldine Atherton, 1972).

11. M. Schmink, 'The Socioeconomic Matrix of Deforestation', in *Population and the Environment: Methods, Cases and Policies*, eds L. Arizpe, P. Stone and D. Majors (New York: Social Science Research Council (SSRC), 1994).

12. C. Jolly, 'Four Theories of Population Change and the Environment', paper presented at the Population Association of America Annual Meeting, Washington, DC, 21–23 March 1991.

13. G. Martine, 'Population, Environment and Development: Key Issues for the End-of-Century Scenario', paper presented at the Workshop on Population Programme Policies: New Directions, organized by UNFPA and NESDB, Chiang Mai, September 1992; and G. Martine, 'Población, Crecimiento y Modelo de Civilización: Dilemas Ambientales del Dessarrollo', in *Población y Ambiente: ¿Nuevas Interrogantes a Viejos Problemas?* eds H. Izazola and S. Lerner (Mexico City: Sociedad Mexicana de Demografia, El Colegio de México, the Population Council, 1993), 49–62.

14. G.C. Gallopin *et al.*, 'Global Impoverishment, Sustainable Development and the Environment', report to IDRC project: Global Impoverishment and Sustainable Development, Ecological Systems Analysis Group, S.C. Bariloche, Rio Negro Argentina, 1988.

15. See, for example, D. Drummond, 'The Limitation of Human Population: A Natural History', *Science* 4178 (1975): 713–21; A.H. Hawley, *Human Ecology* (Chicago: University of Chicago Press, 1986); R. Netting, *Cultural Ecology*, 2nd edn (Prospect Heights, Illinois: Waveland, 1986).

16. See, for example, F. Tudela, ed., *La Moderización Forzada del Trópico: El Caso de Tabasco, Proyecto Integrado del Golfo* (Mexico City: El Colegio de México, 1989), on Mexico.

17. D. Hogan, 'Impact of Population Growth' (see note 8).

18. *Ibid.*; B. Zaba and J. Clarke, eds, *Environment and Population Change* (Liege, Belgium: Ordina Editions, 1994).

19. E. Leff, 'La Interdisciplinaridad en Las Relaciones Población-Ambiente: Hacia un Paradigma de Demografia Ambiental', in *Población y Ambiente: ¿Nuevas Interrogantes a Viejos Problemas?* eds H. Izazola, S. Lerner (Mexico City: Sociedad Mexicana de Demografia, El Colegio de México, the Population Council, 1993), 27–48.

20. See, for example, L. Arizpe, F. Paz and M. Velazques, *Cultura y Cambio Global: Percepciones Sociales sobre la Desforestacion en la Selva Lacandona* (Cuervavaca, Mexico: Centro Regional de Investigaciones Multidisciplinaria (CRIM), 1993); P. Blaikie and H. Brookfield, eds, *Land Degradation and Society* (New York: Metheun, 1987); H. Izazola and C.M. Marquette, 'Mexico City: Pool of Attraction or Pool of Out-Migrants: Urban Environmental Deterioration and Out-Migration by Middle Class Women and their Families from Mexico City since 1985', in *Population and Environment in Industrialized Countries*, eds A. Potrykowska and J. Clarke (Warsaw: Polish Institute of Geography and Spacial Analysis, 1995); G. Ness, W. Drake and S. Brechin, eds, *Population–Environment Dynamics: Ideas and Observations* (Ann Arbor: University of Michigan, 1993); and Schmink, 'Socioeconomic Matrix'.
21. Ehrlich, *Population Bomb*; and Ehrlich and Ehrlich, *Population Explosion, op. cit.*
22. Meadows *et al.*, *Limits to Growth*; and Meadows *et al.*, *Beyond the Limits, op. cit.*
23. See for example, R. Ridker, 'Resource and Environmental Consequences of Population and Economic Growth', in *World Population and Development,* ed. P. Hauser (Syracuse, New York: Syracuse University Press, 1979), 99–123; Simon, *Ultimate Resource*; Simon, *Population Matters*; Lutz, 'Mauritius'; and J. Bongaarts, 'Population Growth and Global Warming', *Population and Development Review* 37 (1992): 289–319.
24. See, for example, Blaikie and Brookfield, *Land Degradation*; C.M. Marquette and R. Bilsborrow, *Population and Environment in Developing Countries: A Literature Survey and Bibliography of Recommendations for Future Research*, ESA/P/WP.123* (New York: United Nations Population Division, 1994).
25. See, for example, Blaikie and Brookfield, *Land Degradation*; H. Jacobson and M. Price, *A Framework for Research on the Human Dimensions of Global Environmental Change* (Geneva: International Social Science Council and UNESCO, 1990); J. Clarke, ed., *Population and Environment* (Paris: CICRED, 1992); R. Bilsborrow and M. Geores, 'Population Change and Agricultural Intensification in Developing Countries', in *Population and Environment: Rethinking the Debate*, eds L. Arizpe, P. Stone and D. Major (Boulder, Colorado and Oxford, England: Westview Press, 1994); L. Arizpe and M. Velazques, 'Population and Society', in *Population and Environment: Rethinking the Debate, ibid.*, Zaba and Clarke, *Population Change*; Marquette and Bilsborrow, *Population and Environment.*
26. See, for example, S. Stonich, 'The Dynamics of Social Processes and Environmental Destruction: A Central American Case Study', *Population and Development Review* 15 (1989): 269–97; P. Stupp and R. Bilsborrow, 'The Effects of Population Growth on Agriculture in Guatemala', paper presented at the Population Association of America (PAA) Annual Meeting, Baltimore, Maryland, 29 March–1 April 1989; S. Harrison, 'Population, Land Use and Deforestation in Costa Rica, 1950–1983', *Working Paper of the Morrison Institute for Population and Resource Studies* Vol. 24 (Stanford, California: Stanford University,

1990); R. Bilsborrow and P. DeLargy, 'Population Growth, Natural Resource Use and Migration in the Third World: The Cases of Guatemala and Sudan', in *Resources, Environment and Population, Supplement to Volume 16, Population and Development Review*, eds K. Davis and M. Bernstam (New York: The Population Council and Oxford University Press, 1991), 125–47; B. DeWalt and S. Stonich, 'Inequality, Population and Forest Destruction in Honduras', paper presented at the International Union for the Scientific Study of Population (IUSSP), Committee on Population, and Associaçao Brasileira de Estudos Populacionais (ABEP) Seminar on Population and Deforestation in the Humid Tropics, Campinas, Brazil, 30 November–3 December 1992; B. DeWalt *et al.*, 'Population, Aquaculture and Environmental Deterioration: The Gulf of Fonseca, Honduras', paper presented at the Rene Dubos Center Forum on Population, Environment and Development, New York Academy of Medicine, New York, 22–23 September 1993.

27. R. Bilsborrow, 'Rural Poverty, Migration and the Environment in Developing Countries', *The World Bank Policy Research Working Papers*, no. WPS 1017 (Washington, DC: the World Bank, 1992).

28. See, for example, R. Rindfuss, S. Walsh and B. Entwisle, 'Competition for Land and Migration', paper presented at the Annual Meeting of the Population Association of America, New Orleans, Louisiana, 9–11 May 1996.

29. F. Zinn, S. Brechin and G. Ness, 'Perceiving Population–Environment Dynamics: Toward an Applied Local-Level Population–Environment Monitoring System', in *Population–Environment Dynamics: Ideas and Observations*, eds G.D. Ness, W. Drake and S. Brechin (Ann Arbor: University of Michigan, 1993), 357–76.

30. See, for example, 'Cultural Survival, "Geomatics: Who Needs It?"' *Cultural Survival Quarterly* 18 (Winter 1995); and P. Poole, 'Indigenous Peoples, Mapping and Biodiversity Conservation: An Analysis of Current Activities and Opportunities for Applying Geomatics Technologies', *Biodiversity Support Program: Peoples and Forest Program Discussion Paper* (Washington, DC: World Wildlife Fund, the Nature Conservancy, and World Resources Institute, 1995).

31. Poole, 'Indigenous Peoples', *op. cit.*

32. J. Clarke and D. Rhind, *Population Data and Global Environmental Change* (Geneva, Switzerland: International Social Science Research Council (ISSC), 1991).

33. See, for example, M. Cruz, 'Population Pressure and Land Degradation in Developing Countries, with Particular Reference to Asia', paper presented at the United Nations Expert Group Meeting on Population, Environment and Development, New York, 20–24 January 1992; Zaba and Clarke, *Population Change, op. cit.*

34. See, for example, Jacobson and Price, *Human Dimensions*; Clarke and Rhind, *Population Data*; Cruz, 'Population Pressure'; Zinn *et al.*, 'Population–Environment Dynamics', *op. cit.*

35. Cruz, 'Population Pressure', *op. cit.*

36. Population Resource Center (PRC), *Meeting the Policy Challenge: Moving*

from *Conflict to Collaboration on the Population–Environment Nexus: Results from the Population and Initiative* (Princeton, New Jersey: PRC, 1992).
37. R. Merton, *Social Theory and Social Structure,* 3rd edn (Glencoe, Illinois: Free Press, 1968).

References

Marquette, C.M. and R. Bilsborrow (1994) 'Commentary on Paul Harrison's Article "Sex and the Single Planet"', *Human Ecology Review* 1 (Summer/Autumn): 245–7.
Meadows, D. (1985) 'The Limits to Growth Revisited', in P.R. Ehrlich and J.P. Holdren (eds), *The Cassandra Conference: Resources and the Human Predicament* (College Station, Texas: Texas A & M University Press), 257–69.
Shaw, P. (1993) 'Book Review of *The Third Revolution: Environment, Population and a Sustainable World* by P. Harrison', *Population and Development Review* 19: 189–92.

3 Population, Environment and Sustainable Development: Global Issues

Jeffrey N. Jordan

Over the past 30 years, the linkages between population, the environment and socioeconomic development have become a growing concern for both the academic research and policy-making communities. In the past, the treatment of these issues has typically been divided along the vertical lines of specialization that are inherent in the research community, development agencies, and national and international policy-making communities, rather than managed in an integrated fashion. However, overcoming the barriers between areas of specialization is possible, and the international policy-making community has shown some openness to addressing sustainable development in a more integrated fashion. This trend is apparent, for example, in documents such as Agenda 21 which was produced by the UN Conference on Environment and Sustainable Development that was held in Rio de Janeiro in 1992, and the Programme of Action which arose out of the International Conference on Population and Development held in Cairo in 1994.

While the volition is evident, there are still several questions that must be answered if this integrated approach is to be effective. First, is an understanding of the linkages between population and the environment in and of itself sufficient to guide policy-formulation and programme-implementation to achieve sustainable development? Secondly, what are the priorities for implementing activities at the international, national and local levels to promote sustainable development in an integrated manner? Finally, how do these considerations play out in specific settings where environmental degradation is clearly linked to population and poverty issues.

In November 1993, the International Academy of the Environment in Geneva convened a roundtable meeting of experts from diverse regions of the world to facilitate a dialogue on these issues. Much of the background for this chapter stems from the preparations for, the dialogue during, and report of this meeting.[1] As a starting point, a clear conceptual framework for understanding the linkages between population and the environment is presented that can guide research and policy-making in this sphere. The framework is than applied to three case studies – desertification in Africa, environmental degradation in the Bay of Bengal, and the problems of small island states in the South Pacific – and an approach to developing policy recommendations is elucidated.

EFFECTIVE INTERNATIONAL POLICY-MAKING

Experience and thoughtful study have shown that population and environment interactions are neither simple nor straightforward. In order to develop policies to foster sustainable development within the context of current and future population and environmental trends, policy-makers must therefore have:

1. a synthesis of information, data and indicators of environmental and population trends; and
2. a clear understanding of the implications of these trends and of actions that can encourage beneficial or mitigate harmful processes.

However, a clear impediment to this synthesis and understanding is the lack of a conceptual framework for understanding the relationship between population and the environment within which policy-makers can work and evaluate implications.

In an effort to fill this gap, Steve Brechin of Princeton University has developed a conceptual framework to provide a basis for further discussion and deliberation, which is shown in Figure 3.1.[2] This framework attempts to capture some of the complexity that is essential for a fuller understanding of the population–environment nexus. We can begin by reflecting on what is meant by the terms population and environment. In this model, population is defined by parametres of size, distribution, flows and sex/age composition, variables which comprise the more or less classic demographic definition of population. Other population parameters such as density, health and population change are treated as the outcome of interactions with other factors.

The environment is perhaps an even broader and more diffuse

Figure 3.1   Population–environment dynamics: a conceptual framework

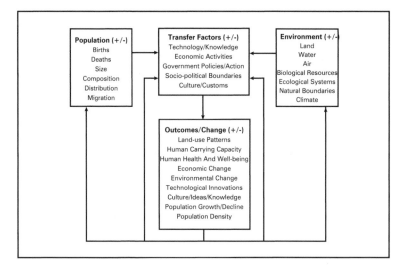

Source: Summary chapter of G.D. Ness, W.D. Drake and S.R. Brechin (eds), *Population–Environment Dynamics: Ideas and Observations*, University of Michigan Press (1993).

concept than population, but it is essential to distinguish precisely what is meant by the term, as this will affect the interactions and outcomes that are identified. In Brechin's framework, environment is defined by variables such as land, water, air and biological resources, as well as climate and ecological systems, with their natural boundaries. Land-use patterns and environmental change are treated as outcomes of the interactions between the environment and other factors.

The two most distinctive elements emerging from this figure, however, are the intermediate or transfer variables, and the concept of distinctive outcomes. The fact that the arrows do not directly connect population and the environment is quite deliberate. This interaction is always mediated by factors such as government policies and action, culture, the state of technology and knowledge, and social or political boundaries. It is the interactions between specific population and environment variables and an array of transfer variables that defines this dynamic and produces specific outcomes such as environmental degradation, population growth or decline, land-use patterns, and the carrying capacity of the land. Finally, these outcomes can in turn affect the nature of the transfer factors, as well as the population and environment conditions, thereby completing the loop in this very complex and dynamic process. Research has found that government policies and actions have an enormous effect in defining population–environment relationships, and so need to be carefully considered and reviewed in order to maintain the health of those relationships.

It is not, however, necessary for researchers to be able to provide policy-makers with absolute knowledge about population and the environment, or their linkages. Many global actions can be taken as preventive measures to ensure that critical thresholds are not breached while scientists seek a better understanding of complex interactions.

A good example of such policy insurance is the passage of global ozone and climate accords. The Montreal Protocol set in motion a global agreement that provided a flexible mechanism with which to respond to the early evidence of the effects of elevated levels of fluorocarbons in the atmosphere. Later research both confirmed and reinforced the threat that this presented for the global environment, as well as the importance of the early response. The flexibility of the protocol allowed governments to begin addressing the issues in a prudent manner that could be readily monitored and evaluated, and the agreement could be altered as the science progressed. It is important for scientists to state their degree of certainty, and their reasonable expectations about what else can be learned. Armed with this type of information, policy-makers can decide whether the partial knowledge available justifies temporary policy inaction in anticipation of further research findings, or the enactment of precautionary policy measures.

INTERVENING VARIABLES

Technological change offers one potential means to loosen the constraints imposed on sustainable development by population and environmental issues. For example, innovations fostered during the green revolution greatly increased the amount of food that could be produced on a fixed amount of agricultural land. However, technology will not always provide solutions before systems are degraded beyond the threshold below which renewal or regeneration is no longer possible. Although there is a clear need to increase the transfer of appropriate advanced technologies throughout the world, policy makers must move away from an excessive reliance on technological solutions.

Another important intervening variable is the behaviour and attitudes of individuals at the local level. To achieve sustainable development, individuals must be committed to change, and generating this commitment will require a worldwide increase in environmental literacy. Environmentally literate populations will keep pressure on their governments for accountability with respect to sustainable development. This literacy must include an understanding of the beliefs and practices of indigenous peoples that serve to preserve the environment and natural resources.

At the same time it is imperative that the poor no longer bear the blame for environmental degradation that occurs in the name of economic necessity. Higher incomes and secure employment would enable the poor to look beyond their immediate subsistence needs. However, it must also be recognized that as families shrink and incomes rise, consumption patterns change in ways that shift environmental impacts from the local to the regional or global levels, particularly as the type and amount of energy used changes. Furthermore, migrants to industrialized countries quickly adopt their prevailing consumption patterns, so emphasis must be placed on both population numbers and consumption patterns.

Formerly, it was feared that significantly altering consumption patterns in the industrialized world would limit the ability of the developing countries to reach their developmental goals. Early OECD models indicated a need for average economic growth rates above 3 per cent per annum in the North to foster development in the South. However, new World Bank/OECD models exhibit less dependence of the South on the North; east and southeast Asia grew substantially despite the last global recession. Nor is the amount of money necessary for generating global improvements in socioeconomic status excessive when considered in the proper context. For example, subsidies to farm production and energy consumption in the North now approach $500 billion per year. In contrast, the estimated cost of supplying clean drinking water globally is only $30 billion.

An important non-economic factor in any effort to address population growth, poverty, consumption and environmental issues is the commitment of political leaders to change. Without political will, well-defined policies and programmes flounder before reaching the implementation stage. Recent success in reducing overall fertility levels in Bangladesh, a country where the status of women and education levels remain low, is largely attributed to strong support for these programmes on the part of the government. While not negating the impact that increasing education and women's status could have on reducing fertility, these gains show the importance of concerted government effort, or, at the least, government facilitation of the efforts of non-governmental organizations.

Finally, intergovernmental and inter-institutional cooperation is paramount for global progress on these issues. Coordinating programmes and research agendas within and between governmental and private organizations and institutions will increase the flow of information and facilitate a more efficient use of resources to approach the obstacles inherent in any effort to create a sustainable world order.

The framework described above can be used to evaluate the population–environment interactions in regions where environmental degradation is serious and population and poverty appear to be contributing factors. This analysis can then provide a stronger basis from which to develop policy recommendations. Here we will consider the cases of desertification in Africa, environmental degradation in the Bay of Bengal, and the problems of small island states in the South Pacific.

DESERTIFICATION IN AFRICA

At a current population growth rate of 3 per cent per year, Africa's population will double in 24 years; by the year 2025 the continent will be home to 1.5 billion people. The total fertility rate in Africa stands at 6.1, and only 16 per cent of married couples are using modern contraceptive methods. Nearly half (45 per cent) of the African population is under the age of 15, and the majority of Africans (70 per cent) still live in rural areas.

The dryland areas of sub-Saharan Africa – 43 per cent of the total area – are home to 66 per cent of the continent's people. Recent estimates suggest that 14 per cent of Africa's total vegetated land surface is moderately, severely or extremely degraded, and an additional 8 per cent is lightly degraded. Much of this degeneration occurs in the form of desertification: the depletion of soil and vegetation within the arid, semi-arid, and sub-humid zones caused largely by human activities.

A principal advantage of traditional agricultural practices in Africa has been the ability of farmers to move around on the land, farming in one place for a number of years, then allowing the land to lie fallow and regain its fertility. However, rapid population growth has largely eliminated this option as farmers are forced to cultivate more intensively, to abandon traditional crop-mixes in favour of cash crops for export, or to migrate to even more marginal lands. Overly-intensive farming and the use of fertilizers and pesticides add to the pressure on soil resources. Changing the crop-mix reduces the ability of local areas to produce enough food to maintain self-sufficiency, while migrating to marginal land increases soil depletion and the competition for water resources. Agricultural encroachment also severely affects Africa's 25 million pastoralists, reducing the availability of grazing land and closing off traditional migration routes.

Meanwhile, between 60 and 95 per cent of sub-Saharan Africa's energy is derived from fuelwood. When over-cutting leaves fuelwood in short supply, animal and crop by-products are burned, preventing their use as much needed fertilizer to compensate for reduced fallow periods. All of these practices – overcultivation, overgrazing, and tree cutting for fuelwood and construction – expose the soil to rain and wind, thus contributing to the process of desertification. Desertification then forces further migrations, completing this vicious cycle. While these migrations temporarily relieve the population pressures on the areas of origin, the migrants often bring inappropriate and destructive land-use practices to the new areas that they occupy.

Some of the problems facing Africa in its battle against desertification are climate-related, such as occasional or chronic drought and the associated desiccation. Other problems are exacerbated by political instability and regional and ethnic conflict, which disrupt markets and increase the flows of migrants and refugees. Land tenure practices are also a factor; a study in Tanzania, for example, has shown that those who have control over their own land also have a heightened awareness of migration and population issues. Thus, it is imperative that efforts to design and implement policies aimed at breaking this cycle of desertification begin with an integrated approach to understanding the complex interactions between problems associated with population, current crop and cattle production practices, energy resources and needs, ecological conditions, land policies and political conditions.

ENVIRONMENTAL DEGRADATION IN THE BAY OF BENGAL

The three countries that border the Bay of Bengal – India, Myanmar and Bangladesh – contain nearly 20 per cent of the world's population. India is the world's second largest country in terms of population, with 897 million inhabitants, while Bangladesh ranks eighth with 114 million, and has the highest population density in the region. Myanmar is much smaller than its neighbours, with a total population of 43 million. Total fertility rates range from 3.9 children per woman in India and Myanmar, to 4.9 in Bangladesh, a marked reduction from the 1970 average of 7.7 children per woman. However, annual population increases remain high, in part due to population momentum; and the age structure of the population (in Bangladesh, nearly 45 per cent of the population is under 15 years old) ensures that populations will continue to grow rapidly despite declining fertility. Each of the three countries will double its population in less than 38 years at present rates of growth.

There are numerous population-linked factors that affect the region surrounding the bay and its varied ecosystems. There are three megacities on the bay with populations over 5 million – Dhaka (Bangladesh), and Madras and Calcutta (India). Bangladesh, at the northern end of the bay, contains the world's largest delta at the confluence of the Ganges, Brahmaputra and Meghna rivers. The Ganges river runs through 114 cities before emptying into the bay, only 21 per cent of which have sewage treatment facilities. In addition to urban sewage and the discharge from over 150 factories, the river carries the runoff from farmlands in which green revolution innovations have increased the content of chemical fertilizers and pesticides.

Past and present population growth continues to place strains upon the varied human and ecological systems in the region. It exacerbates poverty and unemployment problems at a time when there are already few alternative employment prospects; causes rapid urbanization without concomitant infrastructure development; leads to habitat incursion due to the continued expansion of agricultural lands; promotes lowering of water tables and salt water incursions due to extensive irrigation; and increases marine pollution due to oil spillage associated with expanding trade and marine traffic. Overhunting, industrial pollution and habitat incursion are all destroying many forms of plant and animal life, including sea turtles, coral reefs and fish stocks.

The indiscriminate felling of trees to meet local needs for fuelwood and new agricultural lands is destroying forests both inland and along the bay. Inland, the loss of forest cover in conjunction with inappropriate shifting cultivation techniques is leading to soil erosion and the siltation of rivers, exacerbating the yearly flooding in the river basins. The sundurbans, or coastal mangrove forests, serve as a major breeding ground for both fresh and saltwater species of fish, and as a buffer region during the annual cyclones, but these forests are also losing ground to incursions for firewood, cropland and shrimp farms.

While the environmental concerns outlined above are generally linked to actions taken at a local or regional level, the potential rise in sea level as a consequence of global warming presents a particularly damaging scenario for Bangladesh. Potential changes include a 10 per cent increase in rainfall, a seven-fold increase in wet years, submergence of 9 per cent of the country's land surface, and increased flooding on two-thirds of the remaining land.

As is the case for Africa, some of the problems facing the Bay of Bengal region are climate-related, and others relate to policy choices made by governments in the regions. Efforts to control the flows and floods of these rivers have been the subject of frequent negotiations between the governments of India and Bangladesh. Meanwhile, policies to increase preparedness and reduce damage due to cyclones and flooding has begun to help to reduce both human and ecological destruction during these events. And on a global scale, concern for atmospheric conditions through the reduction of greenhouse gas emissions can also improve the region's long-term outlook. Once again, it is clear that as policy-makers search for ways to reduce environmental degradation in the Bay of Bengal, it is critical that an integrated, multi-sectoral approach be taken both to understanding and to resolving the multifaceted problems associated with population growth, irrigated agriculture and shifting cultivation, energy resources and coastal management in the region.

SMALL ISLAND STATES IN THE SOUTH PACIFIC

The 22 nations that comprise the small island states of the South Pacific span a geographic area nearly three times the size of the US or China, but only 2 per cent of this area is land. These island states cover a sea area 24 times as large as the Caribbean island group, making trade, development and intergovernmental cooperation more difficult simply by virtue of the vast distances involved. The islands range in size from tiny atolls to those large enough to support their own rain forests. Common to all of the islands, however, are a thin resource base, aid and remittance-dependent economies, and small domestic markets that are highly vulnerable to external shocks, both natural and economic.

These nations range in size from a population of under 2000 in Pitcairn, to over 3.9 million in Papua-New Guinea, for a total regional population of nearly 6.5 million people. Fertility rates range from a low of 2.7 children per woman in the Northern Marianas, to a high of 7.2 in the Marshall Islands. Rates of population-increase also vary greatly, from 1.4 per cent per annum in Palau, to 3.8 per cent in American Samoa; at this rate the population in American Samoa will double in only 18 years. Contraceptive use in the region is low, ranging from 3 to 15 per cent. There is also substantial migration in the region, both internally and internationally.

Rapid population growth and an underdeveloped human resource base continue to hamper regional development. Furthermore, institutional structures inherited from former colonial rulers are ill-suited to fostering integrated development programmes. Compounding this problem, out-migration to Australia, New Zealand and the US drains off skilled and educated nationals from the human resource base. Meanwhile, local environmental problems are mounting rapidly. Internal migration is contributing to rapid urbanization, while the provision of physical and social infrastructure lags behind. Marine contamination through nutrient pollution is rising, largely due to the dumping of sewage and run-off from cleared land and deforested slopes. Indigenous vegetation is being replaced by introduced plants, while fishing, tourism and mining pressures are endangering coastal and marine habitats through unsustainable fishing practices, industrial pollution, destruction of coral reefs and poor coastal zone management.

The possible rise in sea level and change in climatic patterns as a consequence of global warming also poses a serious long-term threat to the region. A significant increase in sea level would threaten all populations in low-lying coastal areas, while a potential rise in ocean temperatures could increase the frequency and severity of storms in the region.

The nation of Kiribati has served as a focal point for much of the discussion of the problems in this region. The country consists of 30 atolls scattered over 3000 miles, with a population of 72 000. Nearly one-third of the population lives on the southern half of the Jarawa Atoll, whose entire land mass is approximately 35 kilometers long and 200 meters wide. There is rapid urbanization underway amidst a seeming inability on the part of the government to deal with land-management issues. The inhabitants of Jarawa face a broad spectrum of population–environment problems, including poverty, rapid urbanization without adequate services, unemployment and poor marine and coastal management, and a region-wide integrated approach must be pursued to successfully tackle these difficult issues.

AN INTEGRATED APPROACH AND THE SEARCH FOR POLICY SOLUTIONS

The population–environment development nexus is admittedly characterized by complicated interactions in which the chains of causality are often difficult to identify. The impacts of population growth, structure, density and migration are mediated through political, economic, sociocultural, behavioural and institutional factors. They also vary with the environmental conditions and resource base of a particular region during a given time period. Nevertheless, it seems evident from the large body of existing research, as well as from the case studies examined here, that population pressures can exacerbate problems of environmental deterioration and resource depletion and limit the options for sustainable development policies and actions, and that an integrated approach both to understanding and to addressing these problems is necessary.

This chapter attempts to identify an overall approach to developing a rational integrated policy for sustainable development by laying out a conceptual framework for evaluating population–environment linkages, applying it to specific regional issues and finally formulating policy responses. The chart presented in Table 3.1 provides a schema for this process – a method by which to capture

Table 3.1 Framework linking environmental problems, contributing factors, obstacles and actions to address problems

Environmental problems	Contributing factors	Obstacles	Cross-sectoral solutions	Sectoral solutions

Atmosphere
  Climate change
  Ozone layer
Land
  Forest destruction
  Soil erosion
  Desertification
  Arid lands
Water
  Loss of wetlands
  Marine pollution
  Coastal pollution
Biodiversity
  Loss of habitat
Urban pollution
  Wastes
  Urban water supply
  Urban air
	Population
	  Growth rates
	  Density
	  Migration to cities and marginal lands
	Poverty
	Unsustainable farming
	  Land distribution
	  Land tenure
	Consumption patterns
	Market demand from North
	  Primary products
	  Lumber
	  Cattle
	  Cash crops
	Industrial processes
	  Waste
		Low status of women
		  Early marriage
		  Unwanted pregnancies
		  Access to health care
		  Lack of jobs
		  Low education
		Inadequate family planning
		  Son preference
		  High child mortality
		Lifestyles/behavior
		Inappropriate technology
		Lack of political will
		  Loss of confidence in central government
		Institutional
		  Land tenure laws
		  Subsidies
		  Forest logging regulations
		Economic factors
		  Debt burden
		  Terms of trade
		  Market access
			Women and empowerment
			  Education
			  Wealth
			  Jobs-credit
			  Marriage age
			Poverty alleviation
			  Income generation
			Family planning services
			Environmental education
			  Primary schools
			  Media
			Consumption/lifestyles
			  More sustainable
			  Education
			  Taxes
			Local participation and initiatives
			  Community involvement
			  NGOs
			  Insights of indigenous peoples
			Capacity building and infrastructure
			  Financial resources
			  Technical cooperation
				Sustainable agricultures
				  Soil conservation technologies
				  Less use of pesticides and fertilizers
				  Credit and extension services
				Forests
				  Reforestation
				  Legislation
				  Taxes
				  Land reform
				Water
				  Conservation technologies
				  Pricing
				Integrated natural resource conservation programmes
				  Income generation
				  Education
				  Health
				  Industrial processes
				Energy
				  Promotion of alternative sources of energy
				  Promotion of energy efficiency
				  Energy pricing
				Sustainable industrial processes
				  Reduce wastes
				  Product life-cycle
				  Internalize environmental costs

Source: Jeff Jordan, 'Population, Environment and Sustainable Development: Outcomes from the December 1993 Roundtable on Population, Environment and Sustainable Development in the Post-UNCED Period'. Organized by the International Academy of the Environment (IAE) in close collabora-

the issues, their linkages and potential responses. It starts by identifying specific environmental concerns, works through the linkage issues by delineating the contributing factors and the obstacles to successfully addressing the problems, and then moves on to both sectoral and cross-sectoral solutions. It is important to note the range of contributing factors and obstacles that are common to each of the environmental problems listed. In addition, while there are specific sectoral actions that can be taken in an attempt to address environmental problems, these will prove most effective when integrated with cross-sectoral actions. Many of the cross-sectoral solutions are linked to population policy.

Nevertheless, this approach does attempt to mesh both the traditional vertical (that is, divided by sector or discipline) and integrated approaches, because it is necessary to recognize that even when the interconnectedness of issues is understood and the international will to address them in an integrated fashion exists, there are still innumerable forces within both academic and political institutions that will only permit sectoral responses. However, the better the understanding of linkages and the better the outcomes from multisectoral responses, the more likely we are to reach a sustainable future.

Notes

1. The author was both a participant in this meeting, and co-author of the final report. See M. Barberis, R.E. Benedick and J. Jordan (contributors), *Population, Environment and Sustainable Development*, Conference Report for the International Round Table on Population, Environment and Sustainable Development, Geneva, 24–26 November 1993 (Geneva: International Academy of the Environment, with the support of the UN Population Fund, 1994).
2. G.D. Ness, W.D. Drake and S.R. Brechin (eds) *Population–Environment Dynamics: Ideas and Observations* (Ann Arbor: University of Michigan Press, 1993).

4 Population, Affluence or Technology? An Empirical Look at National Carbon Dioxide Production

William R. Moomaw and
D. Mark Tullis

A major debate is currently being waged regarding the causes of environmental problems. Those with a Malthusian perspective argue that environmental problems are ultimately population-driven, and that the rapid growth of population in developing countries is a major cause of local and global environmental deterioration.[1] Others argue that the industrial nations are primarily responsible for environmental damage, and that affluence is the major cause of pollution.[2] In international fora such as the 1992 UN Conference on Environment and Development in Rio, this debate often splits along North–South lines, creating tension during international negotiations. The population–affluence debate continued at the International Conference on Population and Development held in Cairo in September 1994. A third perspective focuses instead on technology, arguing that technological choices are either the cause of, or the solution to, environmental problems.[3]

Although some empirical studies examining local pollution factors were carried out more than 20 years ago,[4] relatively little has been done recently to explore the relative contribution of various factors to global environmental problems, and how they differ among nations. Carbon dioxide (CO_2) emissions have been chosen for this study because the debate over affluence, population and the need for technology transfer has been especially intense during the development of the Climate Convention. In addition, an excellent database exists that records CO_2 emissions since 1950, so comparisons can be made over a four-decade period from 1950 through 1990. The growth of CO_2 emissions during this period thus constitutes an important and instructive policy case that lends itself to empirical testing of the alternative explanations for environmental problems.

METHODOLOGY

Carbon dioxide is released when coal, oil or natural gas is burned for fuel. This gas can trap radiant heat from the earth in the atmosphere, leading to global as well as regional climate change. With available data it is possible to examine the relative contributions of population, affluence and technology (or carbon intensity) to the total CO_2 releases arising from energy production in selected countries. As is customary in the field of climate change, we measure the release of CO_2 in terms of the metric tonnes of carbon in the CO_2.

We utilize two techniques for examining the relationship between CO_2 emissions and population. In the first we present a temporally-connected scatterplot showing the correlation between total national carbon emissions and per capita gross domestic product (GDP). Per capita GDP figures are often equated with wealth, and their increase is the principle measure of economic well-being used by most governments and development agencies. By temporally connecting the values that correlate per capita GDP with CO_2 emissions, we create a development trajectory over time for each country.[5] The purchasing power parity (PPP) measure of GDP developed by Summers and Heston[6] is used, as it provides comparable wealth measures among countries over time in constant, 1985 US dollars. The CO_2 emission data is from the work of Marland *et al.*, from the Oak Ridge National Laboratory,[7] and is given in thousands of metric tonnes of carbon.

The second way in which we explore the relative contributions of population, affluence and technology to national CO_2 emissions is to employ what is called IPAT analysis. In this analysis, which is based upon the work of Ehrlich and Holdren,[8] we utilize the simple identity

Impact = Population × Affluence × Technology

or

I = PAT

The assumption in the IPAT approach is that *each* of the three factors independently drives emissions, and that the fractional increase (or decrease) in each individual factor contributes proportionally to the overall impact. The environmental impact of the population factor is assumed to be simply a multiple of the per capita impact, where the per capita impact is the product of the affluence factor and the technology factor. Affluence is expressed as average GDP per capita, and is really a surrogate for consumption; countries with greater per capita wealth are assumed to consume more per person than those that are poorer. The technology factor is measured as the total environmental impact (tonnes of carbon) divided by the total GDP. This technology factor thus measures the carbon intensity of the economy (tonnes of carbon per dollar of GDP), and implicitly includes the total amount of energy used, fuel choices, the efficiency of fuel use and the structure of the national economy. Economies that choose technologies that produce little pollution, use resources efficiently and have more services and less manufacturing, will have a low 'technology factor' in this analysis. Finally, we do not attempt to define a damage function for CO_2 emissions, but simply assume that climate-change impacts will be proportional to total CO_2 emissions. Thus, for year Y, the relationship between carbon emissions and the three factors then becomes:

$$I(Y) = P(Y) \times A(Y) \times T(Y)$$

where $I(Y)$ is the Impact (tonnes of carbon); $P(Y)$ the Population (number of persons); $A(Y)$ is Affluence ($ GDP/person); and $T(Y)$ is Technology (tonnes of carbon/$ GDP).

To utilize the IPAT equation in analyzing the relative importance of each factor, we create scenarios that hold two of the factors constant at 1950 levels while the third takes on its actual value for each year. We then calculate the hypothetical amount of carbon that would have been released in each year under these assumptions. The actual emissions of CO_2 are then compared with the hypothetical emissions that would have occurred if (1) only population had changed since 1950 – the population effect; (2) only affluence (consumption) had changed – the affluence effect; and (3) only technology had changed – the technology effect.

FINDINGS

From among all of the cases studied, 12 countries have been selected that illustrate different categories of CO_2 emission dependence. The development paths of four of these – the US, China, Poland and Mexico – are shown in Figures 4.1a to 4.4a. We have also separated the contributions to CO_2 emissions into two components: the contribution due to changes in CO_2 production per capita, and that due to changes in population. In the trajectory marked CO_2/capita effect, we have held population constant at 1950 levels and calculated what CO_2 emissions would have been if only carbon intensity per capita had changed as it did over time. For the trajectory marked population effect, the carbon intensity per capita was frozen at 1950 levels, and CO_2 emission rates were calculated as though population was the only factor that changed. The total carbon emissions for any single year are the product of the values of these two factors for that year.

Surprisingly, the population factor is the largest not only for Mexico, but also for the US. Were it not for population growth, US emissions in 1990 would still be near their 1950 levels. In Poland, and even in China, population growth has played only a small role in increasing CO_2 emissions. China's CO_2 emissions have grown 31-fold between 1950 and 1990 (from a very small base), while the country's population has only doubled. The remainder of the increase (15.5-fold) is due to increasing per capita CO_2 production.

These countries have been divided into four categories based upon their degree of industrialization. Contrary to conventional wisdom, economic growth has continued in the industrial countries (ICs) even as CO_2 emissions and the use of fossil fuels have leveled off or decreased. In the transitional industrial countries of Eastern Europe (TICs), emissions have been declining along with per capita GDP due to the recent restructuring of these economies, while most of the newly industrializing countries (NICs) show a strong correlation between rising economic growth and increasing carbon emissions. Some of the least industrialized countries (LICs) exhibit a somewhat more complex, and in some cases chaotic, development path, with little obvious relationship between CO_2 emissions and per capita GDP.

Using the IPAT analysis, we can break down the sources of carbon emissions based on the three contributing factors – population, affluence and technology – and plot them over time. The results are shown for the same four countries in Figures 4.1b to 4.4b. For the US (Figure 4.1b), the growth in total carbon emissions generally reflects the increase in affluence, but if increasing affluence had been the only factor, carbon emissions in 1990 would have

Figure 4.1a   CO$_2$ development plane, USA 1950–1990

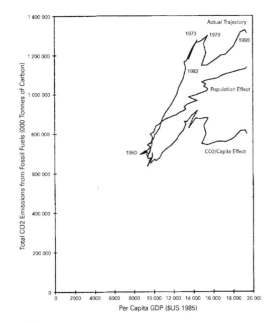

Figure 4.2a   CO$_2$ development plane, China 1950–1990

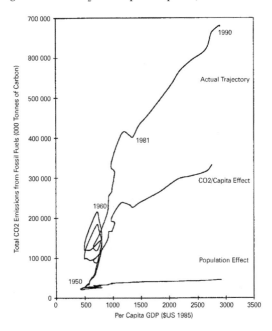

Figure 4.1b CO_2 factors, USA 1950–1990

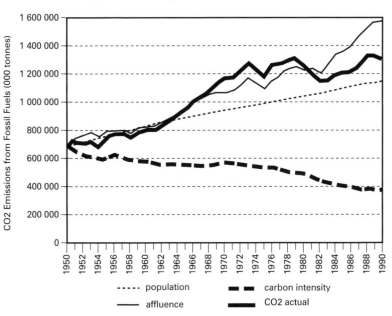

Figure 4.2b CO_2 factors, China 1950–1990

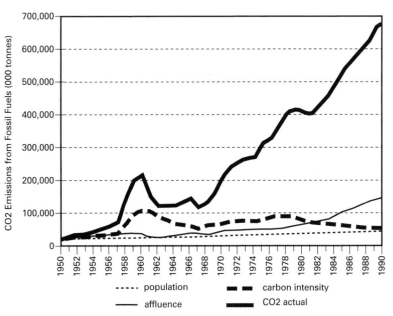

Figure 4.3a   CO₂ development plane, Poland 1950–1990

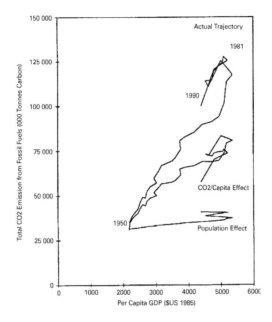

Figure 4.4a   CO₂ development plane, Mexico 1950–1990

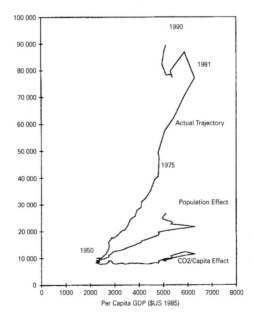

Figure 4.3b CO_2 factors, Poland 1950–1990

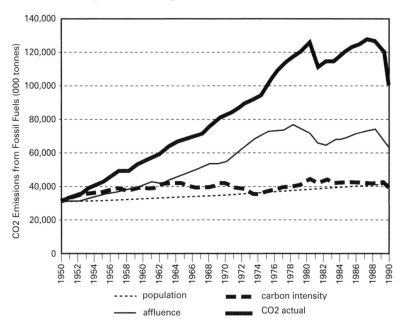

Figure 4.4b CO_2 factors, Mexico 1950–1990

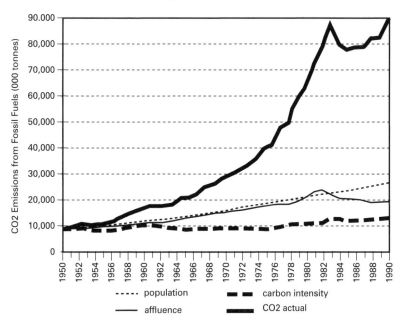

been 258 million tonnes higher than the actual level of 1310 million tonnes. If, on the other hand, population growth had been the only factor, CO_2 emissions would have grown slightly less than actually observed. But the main reason that actual CO_2 emissions only increased by a factor of 1.88 over 41 years is because carbon intensity (the technology factor) decreased to only 51 per cent of its original level during this period. Thus, if population and per capita GDP had remained at 1950 levels, *the US would have produced only half as much CO_2 in 1990 as it did in 1950!*

For China, carbon intensity (the technology factor) dominates the trends in carbon emissions until 1981, but increasing affluence has become the most influential factor since that time. Population growth makes the least contribution throughout this period. In the case of Mexico, on the other hand, aside from a short period of rapidly increasing affluence in the early 1980s, population has always been the dominant factor. In Poland, carbon emissions have tended to rise and fall primarily in response to changing levels of affluence, the factor that has consistently dominated trends in the country, except during the 1950s. Carbon intensity has remained nearly constant in Poland throughout the entire period.

These findings are summarized in Table 4.1 for these four countries plus eight others that are at different positions along their development paths. These 12 countries have been selected from the four different categories listed above (ICs, NICs, TICs and LICs) to illustrate different types of development trajectories. The emphasis is on the major CO_2 emitters, but several other countries are also included for comparison. The countries are listed in order of their CO_2 releases in 1990, and each country's global ranking as an emitter is shown in parentheses.

The next three columns provide data on actual 1990 and 1950 emissions, and list the ratio of 1990 to 1950 levels. The remaining columns are divided into three sets of information for each of the three factors, population, affluence and technology respectively. Each set of three columns indicates (1) the projected emissions levels if the relevant factor alone had changed, while the other two factors had remained constant at their 1950 levels; (2) the ratio between the 1990 and 1950 levels of the relevant factor; and (3) the rank of the relevant factor in terms of its contributions to changes in carbon emissions during this time period (1, 2, or 3).

Actual carbon emissions in long-established industrial countries such as the US and France increased by less than a factor of two

Table 4.1 1990 carbon emissions by individual factors relative to 1950 levels

Country (global emissions rank)	Actual carbon emissions			Population			Affluence			Technology		
	Million tonnes C 1990	Million tonnes C 1950	Ratio 1990: 1950	Million tonnes C 1990	Ratio 1990: 1950	Factor rank	Million tonnes C 1990	Ratio 1990: 1950	Factor rank	Million tonnes C 1990	Ratio 1990: 1950	Factor rank
United States (1)	1 310.341	695.612	1.88	1 137.480	1.64	2	1 567.655	2.25	1	355.570	0.51	3
F. Soviet Union (2)	1 055.499	185.659	5.69	296.511	1.60	2	618.455	3.33	1	198.400	1.07	3
China (3)	678.016	21.667	31.29	44.504	2.05	3	146.414	6.76	1	48.847	2.25	2
Japan (4)	289.288	28.295	10.22	41.773	1.48	2	276.642	9.78	1	20.042	0.71	3
India (6)	183.547	18.272	10.05	42.301	2.32	2	30.590	1.67	3	47.358	2.59	1
Poland (10)	100.040	30.950	3.23	40.762	1.32	2	62.588	2.02	1	37.600	1.21	3
France (11)	97.432	55.668	1.75	75.042	1.35	2	176.906	3.18	1	22.744	0.41	3
Mexico (12)	89.396	8.347	10.71	26.673	3.20	1	19.149	2.29	2	12.194	1.46	3
Brazil (20)	54.379	5.387	10.09	15.176	2.82	2	24.752	4.59	1	4.191	0.78	3
Indonesia (23)*	25.253	2.710	9.32	5.016	1.85	2	10.061	3.71	1	3.675	1.36	3
Philippines (47)	11.752	1.017	11.56	2.993	2.94	1	2.392	2.35	2	1.703	1.67	3
Ghana (87)**	0.899	0.245	3.67	0.670	2.73	1	0.210	0.86	3	0.383	1.56	2

* 1962–90 ** 1955–90.

Source: GDP data from R. Summers and A. Heston, 'The Penn World Table (Mark V): An Expanded Set of International Comparisons, 1950–1988. Quarterly Journal of Economics 106 (1991). CO_2 data from Marland et al., Estimates of CO^2 Emissions from Fossil Fuel Burning and Cement Manufacturing using the UN Energy Statistics and the US Bureau of Mines Cement Manufacturing Data (Oak Ridge, Tennessee: Oak Ridge National Laboratory, 1993) .

during this 41-year period, while in the two TICs on the list, Poland and the former Soviet Union, they increased between three and six-fold. In six countries – Japan, India, Mexico, Brazil, Indonesia and the Philippines – CO_2 emissions increased approximately ten-fold. China saw the largest rise in emissions, experiencing a 31-fold increase, while Ghana exhibited a relatively modest four-fold change.

Population growth makes the greatest contribution to increases in carbon emissions for Mexico, the Philippines and Ghana, and it is the second most important factor for all of the other countries except China, where it is has the least significance. Between 1950 and 1990, population increased by a factor of between 1.3 and 1.9 in Poland, France, Japan, the former Soviet Union, the US and Indonesia. It increased approximately 2.0 to 2.5-fold in China and India during this time, and 2.5 to 3.0-fold in Ghana, Brazil, the Philippines and Mexico.

In eight countries, increasing affluence is the primary factor contributing to increases in CO_2 emissions between 1950 and 1990. This finding was expected in the cases of the US, Japan and France, but it is somewhat surprising to learn that gains in GDP per capita are also the principle factor in the former Soviet Union, Poland, China, Brazil and Indonesia. Affluence is the second most important determinant for Mexico and the Philippines, and the least significant for India and Ghana; Ghana is the only country in the study that has seen its per capita wealth decline during this period, while India's increased only a modest 1.7-fold. The US, Mexico, Poland and the Philippines have slightly more than doubled their per capita wealth, while in the former Soviet Union and Indonesia it has increased more than three-fold. The effects of affluence were most pronounced in Brazil which experienced nearly a five-fold increase, China with a seven-fold rise, and Japan where per capita GDP grew nearly ten-fold between 1950 and 1990.

The technology factor, or increasing carbon intensity, ranks as the first factor only for India, and as the second only in China and Ghana; it is the least significant factor for the other nine countries. India is one of the few countries in the world where carbon intensity continues to increase, suggesting that the country's technological choices are not particularly efficient. Carbon intensity decreased during this period in France, the US, Japan and Brazil to only 40 to 80 per cent of its 1950 level. In the remaining eight countries it increased by factors of between 1.2 and 2.6. The dramatic reductions in carbon intensity in the US and France over this time period substantially offset the effects of increasing affluence and population.

CONCLUSIONS

Our findings reveal a surprising diversity in patterns of CO_2 production among nations. The relative contribution of different factors does not correlate well with the usual ways in which countries are grouped for analysis. The very large growth in China's emissions reflects its very low level of industrialization in 1950 and the massive expansion of its industrial economy. Surprisingly, population growth proved to be the least important factor in the rise in Chinese carbon releases. Japan's rapid increase in CO_2 emissions reflects both its recovery from World War II, and its rise as a preeminent industrial power. The US and the former Soviet Union, on the other hand, have always had energy-intensive economies based upon abundant fossil fuel resources. Population is a 'middle-driver' for CO_2 emissions in most countries, including ICs, TICs and most NICs; it is the principle driver only in the least industrialized countries (LICs) where population growth rates are high.

Our findings clearly show that population growth by itself is an important driver of CO_2 emissions mainly for low-income countries, and that it is rising affluence that correlates most strongly with carbon emissions in most other countries. The debate over what drives pollution therefore needs to be recast to recognize that it is not only the quantity of people, but also the quality of their development choices that is important. Furthermore, the relative significance of the three factors can change over time, a finding that does not appear to have been previously recognized.

Our analysis also demonstrates the crucial role of evolving technology in slowing the growth of CO_2 emissions, regardless of whether population or affluence has been the dominant factor in this increase. This finding, along with the sharp lowering of carbon emissions seen in several countries, suggests the importance of implementing technology-transfer strategies that can offset the emissions increases that will arise from anticipated growth in affluence and population. In the long term it will be necessary to utilize strategies that affect all three factors if we are to succeed in slowing CO_2 emissions. These conclusions should be considered encouraging because they demonstrate that patterns of development are much more dynamic and varied than usually appreciated. Hence, past patterns need not be followed in the future, and a less polluting approach to economic development is possible for all countries.

Notes

1. See, for example, G. Hardin, 'The Tragedy of the Commons', *Science* 162 (1968): 1243–8; P.R. Ehrlich and A.H. Ehrlich, *The Population Explosion* (New York: Simon & Schuster, 1990).
2. W. Howard, 'Man's Population–Environment Crisis', *Natural Resources Lawyer* 4 (1971): 106.
3. See, for example, B. Commoner, M. Corr and P.J. Stamler, *Data on the United States Economy of Relevance to Environmental Problems*, prepared for the Committee on Environmental Alterations (Washington, DC: American Association for the Advancement of Science, 1971); A. Lovins, *Soft Energy Paths: Toward a Durable Peace* (New York: Harper & Row, 1977); and C. Flavin and N. Lenssen, 'Reshaping the Power Industry', in *State of the World 1994* (1994): 61–80.
4. See, for example, B. Commoner *et al.*, 'Data on the US Economy' *op. cit.*; and B. Commoner, M. Corr and P.J. Stamler, 'The Causes of Pollution', *Environment* 13 (1971): 2.
5. W.R. Moomaw and D.M. Tullis, 'Charting Development Paths: A Multi-Country Comparison of Carbon Dioxide Emissions', in *Industrial Ecology and Global Change,* ed. R. Socolow (New York and London: Cambridge University Press, 1994).
6. R. Summers and A. Heston, 'The Penn World Table (Mark V): An Expanded Set of International Comparisons, 1950–1988', *Quarterly Journal of Economics* 106 (1991): 327–68.
7. G. Marland, T.A. Boden, R.C. Griffin, S.F. Huang, P. Kanciruk and T.R. Nelson, *Estimates of CO₂ Emissions from Fossil Fuel Burning and Cement Manufacturing Using the United Nations Energy Statistics and the US Bureau of Mines Cement Manufacturing Data* (Oak Ridge, Tennessee: Oak Ridge National Laboratory, 1993).
8. P.R. Ehrlich and J.P. Holdren, 'Impact on Population Growth', *Science* 171 (1974): 1212–7.

5 A Multivariate Analysis of Farm Household Land-Use and Forest-Clearing Decisions in the Amazon Region of Ecuador[1]

Francisco J. Pichón

As one of the last agricultural frontiers of the humid tropics, Amazonia is the largest area of the world currently undergoing frontier settlement. Much of this movement represents the spontaneous migration of peoples, but governments in the region have also become increasingly interested in opening up and integrating Amazonia into national and international economies in order to increase agricultural production, correct spatial imbalances in the distribution of the population, exploit frontier lands for reasons of national security, and defuse potentially serious political problems associated with the existing agrarian structure, such as landlessness and unemployment.

The upper basin of the Amazon in Ecuador is one such area of frontier settlement. Most of the forest intervention in this region has come at the hands of colonist farmers, who attempt to establish land claims along transport routes that were originally constructed to aid in petroleum exploration and exploitation. These farmers have been squeezed out of their homelands by a variety of factors, including population pressures, pervasive poverty, maldistribution of farmland, lack of inputs for intensive cultivation, lack of non-agrarian livelihood opportunities, and generally inadequate rural development. Marginalized by virtue of their low socioeconomic and political status, they often perceive no way to sustain their families other than by seeking a livelihood in the marginal environments of tropical rainforests.

The growing concern about the negative environmental and social impacts of frontier expansion into the Amazon basin has raised important questions about the viability of small farm colonization as a policy option. Colonist farmers have often failed in frontier regions, losing their lands to larger interests, or abandoning them in the face of increasing degradation of natural resources and land conflicts. There is also growing concern about the socioeconomic and political stratification of frontier society, which appears to be leading to unsustainable farm sizes, second-generation out-migration, and rising unemployment in frontier towns, as well as violence and reliance on illegal crops such as coca. These trends call into question the efficacy of frontier settlement as a natural resource and social development policy-option, and belie the idealistic goal of frontier settlement as a way to bring together 'land without people and people without land'. Latin American governments are therefore struggling to develop policies that can rationalize settlement and development processes on their Amazon frontiers, but reconciling the need to enhance economic productivity with resource protection has proved difficult.

Linkages between demographic processes, agricultural settlement and environmental degradation in frontier regions are extraordinarily complex, and involve a number of disciplines within both the natural and social sciences. Efforts to disaggregate these linkages in Amazonia have therefore encountered a number of difficulties. For example, by focusing within their disciplines, analysts often neglect one or another of the key components in these relationships. In addition, much of the existing data on the region (for example remote sensing data on soil types, vegetation and land-use patterns) is only available at a very high level of aggregation.[2] In the absence of farm-level data, previous studies of farmers' production patterns in Ecuador and elsewhere have only been able to observe general *trends* in land and resource use, without developing an understanding of the socioeconomic and agroecological forces that *underlie* farmers' land-use decisions. Similarly, while forest clearing may be done for a variety of reasons, with very different determinants and consequences, frequently only a *single* indicator of deforestation is used.

The natural resource base must ultimately sustainably provide for the livelihoods of those residing in a given area, so in Ecuador, as in many other parts of the humid tropics, the future of tropical forests hinges on colonists' adoption of more sustainable land-use practices. Any policy- or technology-based effort to alter current production systems must be based on an improved, empirically-based understanding of the ecological, socioeconomic and institutional factors that drive colonists' *farm-level* decisions, and the links between land-use and farmers' welfare. This knowledge is a prerequisite for identifying policy-manipulable 'entry points' or levers to alter farmers' land-use patterns in desirable ways.

Ecuador is a particularly important country for conducting such an investigation for two reasons. The first is its rich biological diversity; the area just east of the Andes – the study site – has been declared by ecologist Norman Myers to be one of the world's ten 'hot spots' in terms of biodiversity.[3] Second, over half of the government's revenues and foreign exchange earnings are derived from petroleum extracted from precisely that part of the Ecuadorian Amazon that is undergoing massive spontaneous colonization. The population of the Amazon region in Ecuador is also growing faster than that in any of the other Amazonian countries, and, correspondingly, its rate of deforestation is also the highest.[4] After reviewing the current state of knowledge regarding the dynamics of resource allocation in frontier environments, this chapter proposes a conceptual framework for understanding colonists' land-allocation behaviour, describes the findings of field research and the analysis of allocation decisions, and draws out their policy implications.

RESOURCE ALLOCATION IN FRONTIER ENVIRONMENTS

The expansion of Latin American agricultural frontiers has given rise to an extensive and varied theoretical debate. These analyses underscore the importance of viewing frontier settlement as an evolutionary process, whose character evolves in sometimes contradictory ways presenting a changing social and natural environmental landscape for migrants. What migrant settlers have in common, however, is a production system characterized by intensive use of family labour and simple agricultural technologies, a strong drive for cattle ownership, and overexploitation of land through continuous incorporation of new areas with little regard for the long-term preservation of the natural resource base. Government policies have also played a significant role in the development of the frontier via provision of subsidized credit, regional fiscal incentives, differential taxation, and a series of other policies that have encouraged inefficient and destructive forms of land-use and forest intervention.

Contrary views have emerged regarding how settlers in the Amazon can advance beyond the level of subsistence without destroying their natural resource base. Although poorly understood, the cropping systems they establish have, in general, not been sustainable under the conditions of the tropical rainforest environment. It is widely believed that farmers, constrained by a 'straitjacket' of natural resources, move through a similar progression of land-use patterns over time.[5] This progression has been described by the World Bank and others as the 'peasant pioneer cycle'.[6] The cycle begins with the occupation of forest lands, requiring some initial deforestation to establish ownership and produce essential food crops. Colonists later clear additional lands for perennial crops as they become more settled. As soil nutrient levels fall, they clear still more land for both perennial and annual crops, leaving the initially deforested areas for pasture and/or fallow. In this way, driven by environmental constraints, survival needs and the absence or unavailability of affordable agricultural technologies that better sustain soil fertility, farmers have little choice but to continue encroaching on the forests.

Many analysts, however, question the prevalence of this generalized pattern of forest clearing, suggesting that increased attention be given to understanding the influence of political, social and economic factors on land-use patterns and technology choices.[7] For example, new research findings suggest that these so-called unsustainable agricultural practices are more sustainable than previously thought, and that frontier settlement is a process of learning-by-doing and adaptation to a new environment.[8] Farm-level data available from these studies indicate that settlers are increasingly finding more sustainable agricultural methods, and they are also improving living standards.[9] It is thus inappropriate to judge the success of colonization efforts while farmers are still in the learning and adapting stage of settlement.

Despite these encouraging findings, rapid turnover and abandonment of degraded agricultural lands continues to be reported. In addition, agricultural colonists continue to be perceived as highly mobile, speculative and uninterested in long-term natural resource development. Analysts have interpreted this instability as evidence that declining crop yields and increasing poverty force settlers to abandon their farms and seek new lands. However, recent studies in the Brazilian Amazon suggest that this trend has less to do with the decline in fertility than with changes in property rights resulting, in part, from unbalanced resource endowments on the frontier.[10] According to these studies, frontier areas are characterized by special circumstances, including abundance of land and relative scarcity of human labour and capital, as well as income levels that are usually insufficient to finance technological changes or investments that preserve the natural resource base. Development and adoption of *long-term* natural resource-conserving systems is therefore an unattractive option, and land turnover and abandonment are consequently high.

Whether settler instability is evidence of loss of agricultural productivity or can be explained by rapid changes in property rights, the available evidence reveals the need for more accurate diagnosis, including distinguishing the relative importance of socioeconomic and agronomic factors in determining the speed and manner of settlement on the frontier. Although seldom empirically documented, the extensive literature on the forest conversion process emphasizes the variability of settler land-use strategies resulting from differences in land access and tenure, labour availability, local infrastructure, soil quality, and other environmental features of the settled region. Smallholder farming in Amazonia is thus much more complicated than the above peasant pioneer cycle or any other forest-to-pasture sequence suggests.

COLONIST LAND-USE DECISIONS: A CONCEPTUAL FRAMEWORK

Land-use strategies reflect colonists' management of land, labour, capital and technology based upon knowledge of agroecological conditions, market opportunities and household consumption demands. Farm management is thus a product of adaptive strategies – often based on diversifying the household's income sources and agricultural activities[11] – aimed at spreading risk, achieving food security and improving the socioeconomic situation of the household in the face of fragile and often changing environmental conditions, as well as uncertainty due to fluctuating harvests and agricultural prices. The strategies selected will in turn have impacts on the prevailing ecological conditions (for example soil quality, erosion, vegetative cover and incidence of pests and disease), thus requiring further adaptations.

Existing models of farm household decision-making have been based on environments characterized by high population densities and land scarcity, conditions that are the opposite of those prevailing in frontier regions such as the Amazon where households typically

Figure 5.1   Conceptual framework to measure effects of
household/farm characteristics, and the broad policy/institutional
environment on land-use options for agricultural colonists in the
Ecuadorian Amazon frontier

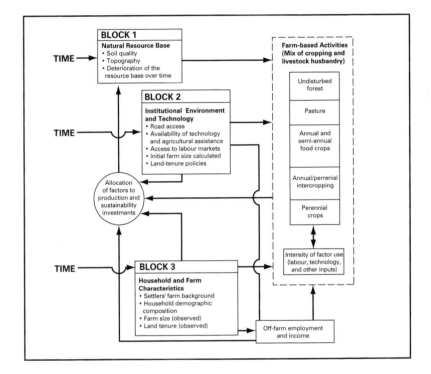

Source: Francisco J. Pichon, 'Agricultural Settlement, Land-use, and Deforestation in the Ecuadorian Amazon Frontier: A Micro-level Analysis
of Colonists' Land Allocation Behaviour', paper presented to ICSE Conference, 1994.

face limited capital and scarce labour. Figure 5.1 depicts a conceptual framework for investigating the effects of household and farm
characteristics and of the broader policy/institutional environment
on the land-use options available to agricultural colonists in the
Ecuadorian Amazon frontier that takes these special circumstances
into account.[12]

As shown in Figure 5.1, the Amazonian colonist is vulnerable to
the dictates of the fragile natural resource base (Block 1) and the
policy/institutional context (Block 2). These two 'filters' tend to be

seen as constraints on the farmer: if access to alternative technologies is limited and the policy environment is not conducive, then the constraints that the natural resource base places on the land-use options available to the colonist (just like a 'straitjacket') are exacerbated. Still, not all farmers wear the same straitjacket, since they control different quantities and qualities of natural resources on their farms, and quality changes (declines in productivity) occur at different rates based in part on a farmer's land-use practices. Moreover, even among farmers with similar natural resource bases, the effects of government policies will vary depending on a given farmer's socioeconomic circumstances. In fact, the effects of both a household's socioeconomic characteristics and the policy environment can *either loosen or tighten* the natural-resource-base straitjacket a farmer inherits with his farm, so it is likely that actual land-use practices will deviate from a common production cycle or sequence that is dictated by natural-resource constraints alone.

SAMPLE POPULATION AND FARM CHARACTERISTICS

The oil-producing northeastern Amazon region of Ecuador, embracing the provinces of Napo and Sucumbios and the river basins of the Napo and Aguarico Rivers, represents a laboratory for the study of smallholder agricultural settlement since it is the part of the Amazon that has received the largest number of in-migrants (Figure 5.2). A detailed survey of settler households in this region was conducted in 1990.[13]

Most land colonization in the region has been spontaneous; no significant settlements have been coordinated or directed by government agencies, though steps have been taken by the government to legalize the land claims of colonists once they have established themselves. Most of the settlers in the sample (nearly 90 per cent) came from other rural areas of Ecuador. Prior to arriving, however, over half of the households had periods of residence in areas other than their places of birth. The vast majority of settlers arrived in the 1970s and 1980s following the oil boom, and the average duration of residence was 10 years.

Education levels among respondents were lower than elsewhere in Ecuador. About one-quarter of household members had completed primary school, but only 10 per cent had formal education beyond this level. Household size averaged 6.6 members, well above the

Figure 5.2   Ecuador and study area

national average of 5.4, and preliminary demographic analysis of the data indicates high fertility and natural population growth rates, as well as high retention of migrant settlers in the study area; most of the children of settlers are remaining in the region when they grow up. Thus, the 'second-generation' population pressures are likely to be substantial, adding to those arising from continuing in-migration to the region.

Over two-thirds of the households sampled had no land of their own before migrating, but there were striking exceptions: over 35 per cent of those who had previously owned land in their areas of origin (11 per cent of the sample) had at least 20 hectares (ha). Thus, although most sample migrants brought little or no assets with them to the Amazon, some colonists had significant personal financial resources. Farmers' land holdings in the study area were generally similar in size, with a mean holding of 46.3 ha, and 80 per cent of farms between 25 and 60 ha. Nearly 84 per cent of the households had only one plot in the region, while the rest had managed to claim or purchase additional plots; no household surveyed owned more than seven plots. More than 93 per cent of sample households had acquired at least a provisional title, and nearly half had a full legal claim to their land. The remaining 7 per cent were primarily claimants of very small plots that had been sold (illegally, strictly speaking) by early settlers to later arrivers. Over 61 per cent of the settlers sampled had obtained their land from previous colonists, while relatively few individuals had inherited their land, and renting and sharecropping arrangements were even rarer.

The average distance between sample farms and the nearest market towns was 28.4 kilometres (km), 22.8 km by unpaved primary and secondary roads, plus 5.6 km by walking and canoeing. Over 50 per cent of the sample plots were accessible only by long walks over muddy jungle paths and/or by canoe. Still, although half of the sample households lacked immediate access to roads, over 85 per cent engaged in market sales. However, only 12 per cent, mainly those next to a road, engaged in timber sales, even though most farms had trees of significant commercial value.

Settler land-use systems involved both new local patterns, as well as adaptations of previous practices from areas of origin. For example, although most settler households in the study area grow coffee, less than 20 per cent had grown coffee in their areas of origin. A polyculture system has evolved in the region based on 'slash-and-mulch' cultivation, which utilizes annuals such as corn and rice, semi-perennials such as plantains, bananas, manioc and yucca, and tree crops such as coffee, cacao, *guaba*, and *achiote*. The slash-and-mulch practice is distinguished from the more commonly known slash-and-burn frontier cultivation system because the felled vegetation is not burned – the high and seasonally continuous precipitation in this part of the Ecuadorian Amazon simply precludes it.[14]

Mean land-use patterns for the farms surveyed included 6 ha (or 16 per cent of the farm area) planted in perennial crops, 2 ha (5 per cent) in food crops, 10 ha (22 per cent) in pasture, and 28 ha

(57 per cent) that remained in undisturbed forest; but the share of cropped area dedicated to each crop varied substantially across farmers. Coffee and cacao were the main cash crops, cultivated by 95 and 36 per cent of the surveyed households respectively, while plantains, manioc, corn and rice were the main subsistence crops. Nearly 57 per cent of the households owned some cattle; herd size averaged 12 head for cattle owners, and seven for all farms sampled, though holdings ranged from 1 to 422 head. Over 45 per cent of the respondents said that, given the chance, they would undoubtedly switch from agriculture to some kind of specialized cattle raising to improve their welfare.

There is little geographic variation in climate across the region studied – the entire area is warm, humid and fairly cloudy – but land quality varies widely. Few settlers complained about soil quality, supporting the belief that superior soils (inceptisols) prevail in this part of the Amazon, but two-thirds of the settlers reported observing declines in crop yields during their time of settlement in the Amazon.

Over two-thirds of the farmers in the study area get along without technical assistance. During the year prior to the survey, 56 per cent of the households had used modern inputs, and 60 per cent employed hired labour. Only 18 per cent – invariably those who could post sufficient collateral in the form of land titles or fixed assets – had ever received credit, and over two-thirds of these recipients had significant debts at the time of the survey. Credit was used mainly for cattle and pasture expansion. Nearly 35 per cent of the households in the survey had a family member employed off the farm in the previous year.

MICRO-LEVEL ANALYSIS OF COLONIST LAND-USE DECISIONS

To investigate how the natural resource base, the institutional environment and household and farm characteristics affect colonists' land-allocation decisions, a set of land-use share regression equations was estimated. Based on the survey results described above, land-use is classified into four mutually exclusive categories – undisturbed forest, annual and semiannual food crops, perennial (tree) crops, and pasture (including fallow land, or *rastrojo*) – and the dependent variables are the percentages of land area allocated to each of these uses. The total land area available to a given farmer is constant, so increasing the allocation for one type of use will require equal decreases in one or more of the others.

All of the explanatory variables in each of the three blocks (see Figure 5.1) are included in the four estimation functions. The natural resource block includes soil quality and deterioration of the resource base with use over time, proxied by the duration of farm settlement. The institutional environment block includes access to and quality of infrastructure, access to technology and agricultural assistance, indebtedness and the use of modern inputs. The household and farm characteristics block includes the farming background and education of the household head, the availability of family labour, hiring in of wage labour, off-farm employment, land tenure and farm size. The relevance of each of these explanatory variables has been justified in the literature, as has the expected sign of the effect in most cases, although in a few cases the expected signs of the effects are not obvious. Most of the statistically significant effects observed were in the anticipated directions, and no sign reversals or significant differences in the magnitudes of the estimated coefficients were observed, so there were no surprises in the results.

Farm-to-farm differences in the natural resource base (the 'straitjacket' within which farmers have to work) did affect land-use, confirming the importance of ecological differences when farmers lack access to technology and other options to alter the productivity of the resource base in the absence of policy incentives or financial resources. For example, farms with more fertile, black organic soils cleared more land and, as expected, allocated larger land areas to both perennial and food crops; while farms with steep, hilly terrain kept more land in forest, as difficult terrain appeared to slow down clearing for agricultural and cattle-raising activities. The link between higher soil quality and greater deforestation contradicts the common view, which argues that excessive deforestation in the Amazon is evidence that the region is simply agronomically unsuitable for agriculture. These results instead suggest that a resource base ill-suited for agricultural production in fact retards the process of forest intervention. These results are of particular importance given that relatively good soils and flat lands are common in the study area.

Ecological factors do not, however, tell the whole story. After controlling for differences in the quality of the natural resources base, the analyses indicate that plots opened earlier had greater proportions of deforested area, but there was no indication that farmers follow a common 'cycle' of land-use over time. Nevertheless, the data show that during the first years of forest intervention, initial crop-planting decisions are dominated by food security concerns (that is, farmers grew more annual and semiannual food crops in newly-cleared lands), while for some farm households the additional land cleared over time tended to be allocated primarily to pasture and other land-using activities. Clearly duration of settlement does play an important role in farm land-use, although it is difficult to discern how much of this effect is due to soil deterioration, and how much to other factors not controlled for in the model that also change with time.

Accessibility to markets and quality of infrastructure significantly affect land-allocation decisions. For example, the greater the distance from the nearest market on a primary or secondary road, and the greater the distance that can only be transited by foot or canoe, the more likely it was that land remained forested. Moreover, the effects of walking (or canoeing) distance alone on land-use were at least twice those of the total distance from markets, indicating that the *inaccessibility* of the plot is more important than just the *distance* on a primary or secondary road in determining the extent of deforestation (probably a result of the much greater drudgery involved in transporting crops by foot or canoe through the jungle). Farms closer to markets and/or nearer a transport road were also more likely to grow pasture and cash crops because of the ease of directly marketing the produce. Surprisingly, this relationship was also observed for annual and semiannual food crops, which were also less likely to be grown on physically inaccessible plots, despite the expectation that remoteness would require greater self-sufficiency and thus be associated with more production of basic food crops.

Farmers who owned larger cattle herds left a smaller share of land in forest and dedicated a larger share to pasture. Since most debt was accumulated to expand cattle herds, the amount of outstanding debt also had a positive relationship to area under pasture (although not to total deforested area). Meanwhile, availability of technical assistance and expenditures on modern inputs did not significantly alter land-use patterns, although the estimated impacts of these variables must be viewed with caution since so few farmers in the sample received agricultural assistance or used modern inputs. Low expenditures on land-improving inputs and labour are consistent with frontier perceptions that it is simply cheaper to indiscriminately exploit the agricultural resource base and then move on than to invest in the sustainability of the productive base.

Contrary to expectations, the agroecological zone of origin of settlers did not influence overall levels of farm deforestation, nor did it appear to affect the share of farm area under pasture, although it did have some influence on crop selection. The educational achievement of the head of the household, meanwhile, had a surprising positive effect on forest clearing, which may be interpreted as evidence that more education stimulates the desire to increase household production and consumption so as to improve living standards. Educational efforts can improve farmers' awareness of the value of preserving some of the land in forest, but even so, the results of such programmes have often been disappointing. Improving the flow of information to a decision-maker does not necessarily increase his or her capacity to act on it.

Not surprisingly, both family labour and wage labour have stronger effects on land-allocation decisions than those of most other variables examined (with the exceptions of distance, title and farm size), confirming the assumption that labour is the limiting factor in production. Households with fewer members working on the farm were more likely to keep larger areas in forest, while greater availability of family labour was associated with more land planted in food and especially perennial crops, although it had little effect on the share of farm area allocated to pasture, probably due to the low labour intensity of cattle ranching. The effects of employment of wage labour were similar, but with significantly greater magnitudes. Average expenditures on wage labour were also double those on modern inputs, suggesting that settlers tend to rely more on labour than purchased inputs to increase productivity.

The availability of off-farm employment in the frontier area is also a potentially important policy variable, since households in which members worked more time off-farm converted less of their forest to perennial crops and pasture. This suggests that diversifying income via off-farm activities can relax the economic pressures that force farmers to clear large areas of forest in order to support their families. Off-farm employment can also generate resources that can be allocated for productivity-improving technologies and/ or investments in the sustainability of the natural resource base.

Security of land tenure also appears to have a significant influence on land-allocation decisions. Farm households that had some title to their land (either definite or provisional) converted less forest to agricultural land, and had smaller shares of farm area cultivated in perennial and food crops and pasture, than households without formal tenure. However, evidence from maturing frontiers in the Brazilian Amazon and Central American regions have indicated that the introduction of improved land-titling by itself can produce a climate that favours land speculation and the consolidation of large landholdings by those with enough capital to buy out small settlers.[15]

As expected, smaller plot size generally led to more intensive use of the land, and thus more forest clearing for annual and perennial crop production. However, while families with larger plots cleared smaller proportions of forest on average, they were more likely to be involved in land-intensive cattle farming; the average area in pasture increased with farm size from three ha on 10–24 ha farms, to 44 ha on farms of 100 ha or more. Cattle ranching is also a labour-saving activity, demanding considerably less labour per unit of area than raising crops. Finally, households owning more land outside the sample area converted less forests on their *sample* plots to agricultural use.

POLICY IMPLICATIONS

It is useful to consider the policy recommendations that arise from these findings at two levels. First, policies to minimize deforestation by existing settler families should focus on stabilizing land-use in the region. Second, policies are needed that address the root causes of the in-migration to the Amazon; that is, the factors underlying the out-migration of the Amazon settlers from their areas of origin.

Stabilization of Farm-Level Land-Use

The economics of cheap, easily accessible land, combined with low population density and lack of capital inputs, encourages unsustainable use of the agricultural resource base, especially via land-using, labour-saving cattle ranching; agriculture is expanded not through intensification, but primarily through extensification. Fortunately, resource endowments are not only accidents of geography, but also products of the market and institutions, so the conditions that affect the profitability of innovations can be changed. The mobility of capital and labour can be enhanced, land-can be enclosed or otherwise regulated, road construction can be more strategically planned, and the legal and institutional framework can be reevaluated to promote more intensive, land-saving agricultural practices in the region. If extensive land-use systems are encouraged by the abundance of cheap land in frontier areas, then this trend can be countered by avoiding government policies that increase the supply of new land, such as opening new roads, allocating unnecessarily large land holdings to settlers, and neglecting to define and enforce clear property rights.

Roads are a fundamental determinant of settlement. Intensification of the road network can increase the price of adjacent land by reducing transport costs and increasing farm profitability, but, at the same time, expansion of the road network can reduce the price of currently accessible land by putting massive quantities of new land on the market. Without major reforms in the places where migration flows originate, limiting the construction of new roads intended to further expand the agricultural frontier may be the only way to reduce in-migration to the Amazon region.

Road construction is a particularly sensitive issue in the Ecuadorian Amazon, where newly built oil-exploration and service roads serve as locational determinants for further influxes of spontaneous settlers, because petroleum extracted from this region is a main component of the country's energy balance and a critical source of economic growth, foreign exchange earnings and government revenues. Nevertheless, if the Ecuadorian government is to rationalize the exploitation of forest and land resources in its Amazon frontier, the most direct policy means is by minimizing new road construction in the region. Government policy should therefore ensure that new road construction is designed to intensify the use of existing accessible land, while a moratorium should be imposed on road projects that open new frontier lands, at least until the overall economic conditions improve enough to alleviate the 'push factors' of migration.

Colonists in the Ecuadorian Amazon do not fit the mould of traditional subsistence farmers, who may limit the area they cultivate once their agricultural production satisfies the nutritional needs of their families. Rather, the areas cleared and planted are limited not by humble ambitions, but by the amount of land, labour and capital farmers have available to expand their agricultural activities.

Land policies should be designed to reduce the size of settlers' land claims to plots that are just large enough to occupy the available labour of each household when using technically recommended farming methods, and farmers should be provided with more capital and technology to expand production at the *intensive* margin. Limiting the size of settlers' plots to something closer to the economically optimal scale will redress the current imbalance of factor proportions that encourages land-intensive, low-labour options such as cattle ranching.

Restricting the area that can potentially be claimed for expansion of agricultural settlement can be achieved through creation of national parks, demarcation of Indian land, creation of extractive reserves, and delimitation of areas that because of inadequate soils and the fragility of the ecosystem should be declared off-limits to settlement. This will require that the region be mapped and agriculturally zoned, and that this is translated into regulations concerning the spatial distribution of production activities and non-production uses.[16] Nevertheless, policy-makers and environmental organizations must also be realistic about the political sustainability of reform measures intended to reduce rural people's access to resources. There is a risk that policies aimed at reducing the ecological impacts of settlers' production activities will take precedence over policies aimed at making their lives better. Attempts to reduce or stop tropical deforestation in frontier regions by fiat must therefore also take account of the factors that determine the ability and willingness of the populations living in or near forested lands to maintain the natural resource base in these areas.[17]

Land tenure is also a central issue underlying settlement and deforestation, especially given farmers' dependence upon the availability of cheap land as the basis for their production systems. Other things being equal, farmers who hold land titles deforest less land than those without title. The combination of insecure tenure and the availability of cheap land encourages settlers to minimize the costs of occupation by turning to premature deforestation, cattle ranching, and other activities that occupy the cleared land at low cost. Land-tenure insecurity thus causes negative environmental effects whenever it leads farmers to forgo investment in the agricultural resource base that they might otherwise have undertaken.

In the absence of a well-articulated smallholder settlement policy, however, the introduction of improved tenure has sometimes increased speculative holding by later, better-endowed, migrants to the region, or by outsiders. The first colonists at the frontier are likely to be people with low opportunity costs who, in the absence of specific settlement policies, can expect to sell out to those arriving later. The late arrivals generally have superior access to credit and government services, making it relatively easy for them to bid the earlier settlers off the land. Land policy should therefore promptly allocate property rights to the initial settler occupants. Granting formal title to early settlers will improve their access to credit (as titled land becomes collateral), thereby reducing the possibility that differential discount rates alone will lead to buyouts, land consolidations and continued expansion of the agricultural frontier.

Governments can also promote farm stability during the initial stages of settlement by providing early colonists with farm-level agricultural research and extension, rural banking, crop marketing, education, health care, recreation, transportation, and other community services that improve human capital and productive capacity in the frontier areas. Enhancing the quality of life in the area of settlement can do much to encourage small farmers to intensify land-use, seek alternative sources of off-farm employment, and invest in the conservation and improvement of their holdings.

Another obvious point of intervention for achieving farm stability in the region is the introduction of more intensive, land-saving agricultural methods. Although still in their infancy, attempts are being made to develop new land-use methods for intensification or modification that are commercially viable and thus attractive to settlers. Many such efforts are based on the modification of traditional systems that have had the benefit of centuries of trial-and-error adaptations by long-term residents of tropical forests who are accustomed to the exigencies of the environment. However, it should be remembered that farmers in frontier regions seek much higher levels of economic return than those obtained using the traditional practices of native peoples whose objectives focused on subsistence.

The challenge for scientists is thus to develop land-use options that are not only agriculturally sustainable but also socially feasible and relatively unrestrained by the limited commodity markets and input sources. Agronomic research has revealed both the appropriateness and the limitations of settler agriculture, and has led to a growing interest in agroforestry and other technological innovations based on permanent or tree crops that are better adapted to the soil properties and productive characteristics of tropical environments.

However, while less destructive than ranching, these new land-use systems have thus far accomplished little in terms of solving problems such as labour scarcity, insecurity of tenure, lack of alternative sources of off-farm employment, and other factors that are more directly responsible for the inability of settlers to establish sustainable forms of land-use. New systems are of little use if households lack the labour to implement recommendations, or if insecurity of tenure makes investing in land-improving inputs too risky to be practical. Moreover, even if successful it should not be presumed that small-scale experimental systems can be easily expanded to widespread forms of land-use in frontier regions. Preoccupation with experimental or *model* systems reflects a long-standing bias towards technological solutions which cannot substitute for confronting the need for broader structural and policy changes.[18]

The Need for an Expanded Approach

These proposals for improving the capabilities of smallholders to manage their resources in frontier environments can help stabilize small-scale agriculture in already occupied areas. However, the agricultural development of the region can hardly be construed as a solution for the serious structural problems that lie beyond Amazonia's borders. Although new and more productive farming systems are clearly needed for the marginal and easily degradable lands of the Amazon, it is wishful thinking to believe that scientists can develop – and farmers readily adopt – sustainable systems quickly enough to stabilize existing farming populations, and also meet the needs of the continuously growing population over future decades.

Much of the decline in the Amazon forests reflects a failure of development policies in general in Ecuador; that is, a failure to tackle problems at the source, particularly the phenomenon of migration of farmers due to the myriad pressures that impel them to move to the frontier. This means that the remedial measures must embrace a whole spectrum of initiatives that relate to the agricultural migrant, addressing the meta-problems of pervasive poverty, population growth and landlessness in the areas of origin. Problems of surplus population and unemployment may be more cheaply and effectively dealt with in their place of origin; what happens in the settled agricultural areas of the country is at least as important as what is done in the marginal lands of the frontier when it comes to reducing resource degradation.

Notes

1. An earlier version of this chapter was presented at the 48th Congress of Americanists: Threatened Peoples and Environments in the Americas, Stockholm/Uppsala, Sweden, 4–9 July 1994. The World Wildlife Fund, the National Science Foundation, the Compton Foundation, the Pew Charitable Trust, the Inter-American Foundation, the Lindbergh Foundation, and the Carolina Population Center at the University of North Carolina at Chapel Hill all provided valuable support for this work. A significantly reviewed version including tables was published in *Economic Development and Cultural Change*, vol. 45, no. 4 July 1997, pp. 707–44 © 1997 by the University of Chicago. All rights reserved 0013-0079/97/4504-0002. The findings, interpretations and conclusions of the author should not be attributed to the World Bank.
2. D. Southgate and M. Whitaker, *Economic Progress and the Environment: One Developing Country's Policy Crisis* (New York: Oxford University Press, 1994).
3. N. Myers, 'Threatened Biotas: Hotspots in Tropical Forests', *The Environmentalist* 8 (1988): 1–20.
4. D. Southgate, 'Policies Contributing to Agricultural Colonization of Latin America's Tropical Forests', in *Managing the World's Tropical Forests: Looking for Balance between Conservation and Development*, ed. I. Sharma (Dubuque: Kendall/Hunt Publishing Co., 1992).
5. A. Oberai (ed.), *Land Settlement Policies and Population Redistribution in Developing Countries: Achievements, Problems and Prospects* (New York: Praeger, 1988). See also J. Hicks, H. Daly, S. Davis and M. Lourdes, *Ecuador: Development Issues and Options for the Amazon Region* (Washington, DC: World Bank, 1990).
6. World Bank, 'Brazil: An Analysis of Environmental Problems in the Amazon', Internal Discussion Paper no. 9104-BR (Washington, DC: World Bank, Latin America and Caribbean Region, 1992).
7. See, J. Collins, 'Smallholder Settlement of Tropical South America: The Social Causes of Ecological Destruction', *Human Organization* 45 (1986): 1–10.
8. See, for example, E. Morán, 'Adaptation and Maladaptation in Newly Settled Areas', in *The Human Ecology of Tropical Land Settlement in Latin America*, eds Schumann and Partridge (Boulder: Westview Press, 1989); and L. Ozorio de Almeida, 'Deforestation and Turnover in Amazon Colonization', unpublished report (Washington, DC: World Bank, 1992).
9. Food and Agriculture Organization/United Nations Development Program/Ministerio da Agricultura e Reforma Agraria (FAO/UNDP/MARA), 1992; and Ozorio de Almeida, 1992.
10. For an extensive discussion of this work, see R. Schneider, 'Government and the Economy on the Amazon Frontier,' Regional Studies Program, Report no. 34 (World Bank, Latin America and the Caribbean Technical Department, Washington, DC, 1994).
11. O. Arguello, 'Estrategias de Supervivencia' (Survival Strategies), *Demografía y Economía* 15 (1981): 265–311; and M. Todaro, *Economic Development in the Third World* (New York: Longman, 1989).

12. For a more detailed explanation of this conceptual framework, see F. Pichón, 'Settler Agriculture and the Dynamics of Resource Allocation in Frontier Environments', *Human Ecology*, Vol. 24, No. 3, 1996; and F. Pichón, 'Land-Use Strategies in the Amazon Frontier:Farm-level Evidence from Ecuador,' *Human Organization*, vol. 55, No. 4, 1996.
13. See F. Pichón, 'Settler Households and Land-Use Strategies in the Amazon Frontier: Farm-Level Evidence from Ecuador,' *World Development*, vol. 25, No. 1, 1997.
14 F. Pichón, 'The Forest Conversion Process: A Discussion of the Sustainability of Predominant Land Uses Associated with Frontier Expansion in the Amazon', *Agriculture and Human Values*, vol. 13, no. 1, 1996.
15. See J. Collins and M. Painter, 'Settlement and Deforestation in Central America: A Discussion of Development Issues', *Cooperative Agreement on Human Settlements and Natural Resource Systems Analysis* (Worcester, Massachusetts: Clark University; Binghamton, New York: Institute for Development Anthropology, 1986).
16. See E. Morán, 'Monitoring Fertility Degradation of Agricultural Lands in the Lowland Tropics', in *Lands at Risk in the Third World*, eds P. Little and M. Horowitz (Boulder: Westview Press, 1987).
17. See, for example, W. Thiesenhusen, 'Implications of Rural Land Tenure Systems for the Environmental Debate: Three Scenarios', unpublished manuscript, Land Tenure Center, University of Wisconsin, Madison, 1991; and S. Bunker, *Underdeveloping the Amazon* (Urbana: University of Illinois Press, 1985).
18. See, for example, P. Fearnside, 'Rethinking Continuous Cultivation in Amazonia', *BioScience* 37 (1987): 209–14.

6 Population–Environment Dynamics in Lahat: A Case-Study of Deforestation in a Regency of South Sumatra Province, Indonesia[1]

Laurel Heydir

The World Bank estimates that Indonesia loses about 900 000 hectares (ha) (0.63 per cent) of its tropical forests each year,[2] and according to World Resources Institute calculations, if the entire area burned by the catastrophic mid-1980s fire on Kalimantan Island is included, the annual deforestation rate reached 1.3 million ha.[3] The country ranks third globally in terms of total annual forest loss.[4] The principal forces behind tropical deforestation are thought to be agricultural expansion and unsustainable commercial logging, activities that are usually traceable to increasing population and the failure of government policies.[5] In addition, West, Brechin and others identify common problems that affect parks and other protected areas throughout the world, including industrial pollution, excessive tourism, shrinking or non-existent conservation budgets, land fragmentation, economic development pressures and growing rural populations, as well as the technologies and cultural practices that are employed in forest use.[6]

In Indonesia, cultivation by rural farmers is a substantial cause of deforestation. Repetto and Gillis calculate that about 50 per cent of the country's deforestation is caused by slash-and-burn farming, 40 per cent is due to the government's resettlement programmes, and the remaining 10 per cent is caused by commercial logging;[7] while the World Bank estimates that 56 per cent of Indonesia's

Table 6.1   Sources of Indonesia's deforestation

Source	Best estimate (ha)	Range (ha)
Logging and fire loss	180 000	150 000–250 000
Development projects	250 000	200 000–300 000
Smallholder conversion	500 000	350 000–650 000
Total	930 000	700 000–1 200 000

Source: World Bank, Indonesia: Forest, Land, and Water [Issues in Sustainable Development], 1989: xvii.

deforestation is caused by smallholder conversion, 28 per cent by development projects, and the remainder by logging and fire loss.[8] These figures suggest that learning about rural farmers' behaviour is essential to understanding Indonesia's deforestation (see Table 6.1).

A number of international projects in Indonesia have investigated population growth and migration trends, but unfortunately these projects have not generally extended to a study of the effects of population changes on deforestation in specific forest sites. Even in a global context, only a few empirical studies have been conducted on the relationship between people and protected areas, and on the dynamics of farmer encroachment. In order to rectify this problem, this field study was conducted in Lahat regency, a highland regency located in the extreme western part of South Sumatra Province (see Figure 6.1), a rich agricultural region and a major centre for coffee production. The site was chosen because records from the Lahat Office of Forestry indicate that, compared to other regencies within the province, Lahat has the highest rate of deforestation and the largest number of forest farmers, and because Lahat receives considerable government attention due to a proposed relocation project for forest farmers in the regency.

This study starts with the question: 'how and why do small-scale coffee farmers deforest large portions of established protected forests within Lahat?' The study is not intended to examine a particular model of population–environment relationships, nor to defend a particular grand theory. It follows the tradition of grounded research aimed at allowing theory to emerge from the subject under observation. Primary data for the study includes direct observations, and in-depth structured interviews with farmers and key informants, including government officials (local administrators and forestry officers) in provincial, regency and sub-district levels, and (informal) community leaders. Secondary data were obtained from

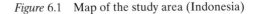

Figure 6.1   Map of the study area (Indonesia)

Source: Modified from Brechin *et al.*, in Ness, Drake and Brechin, eds, *Population Environment Dynamics: Ideas and Observations* (Ann Arbor, MI: University of Michigan Press, 1993) p. 228.

government agencies, the association of South Sumatran coffee exporters, and from satellite images of Lahat's forests which helped clarify the field observations.

According to the latest version of the national forest classification system, the *Tata Guna Hutan Kesepakatan* (TGHK),[9] Indonesia's forests cover nearly 144 million ha, or 71 per cent of the country's total land area. However, not all of this area is, in fact, still forested. According to some estimates, as much as 31 per cent of the land within the designated boundaries is no longer forested. The Forestry Department estimates that about one million families practice shifting cultivation on 7.3 million ha of the country's forest lands.[10]

The Lahat regency administers an area of 701 238 ha, of which about 41 per cent, or 290 600 ha, are designated as TGHK forest that is technically under forest management.[11] Based on the TGHK

Table 6.2   TGHK-classification of forest lands in
Southern Sumatra and Lahat Regency

Forest classification	Southern Sumatra		Lahat Regency	
	Area (ha)	*% of total*	*Area (ha)*	*% of total*
Conservation forests	871 550	17	79 500	27
Protection forests	774 602	15	149 600	52
Limited production forests	333 000	6	21 750	8
Regular production forests	2 124 000	41	39 750	14
Conversion forests	1 111 500	21	0	0
Total	5 214 652	100	290 600	101*

* Does not add to 100 because of rounding.

Source: Bappeda Lahat, 1989: 132 jand Surat Keptusan Mentri Kehutan,
No. 401/Kpts-II/1986.

classification, 79 per cent (229 100 ha) of Lahat's forest is designated as permanent forest, and the remaining 21 per cent is set
aside as production forest; none of the forest can be converted for
agricultural uses (see Table 6.2). Given that arable land within village areas is scarce, the implementation of this forestry policy contributes to land pressure. This is especially true within Lahat regency
where farmers make up 73 per cent of the population.[12]

Field observations reveal several environmental consequences of
deforestation. Deforestation of mountain slopes has exposed soil
to erosion from run-off during the rainy season, leading to land-
slides and floods after heavy rains that have occasionally killed villagers. Erosion clogs the natural waterways and rapidly silts up the
Musi River, which adversely affects both commercial water traffic
and fisheries. In addition, as more water is lost to rapid run-off,
river flows become more irregular. Thus, the village farmers at the
foot of the mountains now face flooding during the rainy season
and water shortages during the dry season, problems that were
previously unknown. Irregular water flow also disrupts the traditional irrigation systems used for rice cultivation.[13]

THE IMPACTS OF FOREST MANAGEMENT POLICIES

The people of South Sumatra live in communes, called *margas*, which
have been the basic native political unit for centuries. A *marga* occu-

pies a certain area of land, known as its *adat* (customary) territory, and forests that stand within the *marga* boundary are part of its domain. These *rimba marga* (*marga*-forests) are used as a community resource for activities such as collecting honey and gathering firewood, and they are physically protected by *marga* members from use by non-*marga* members. *Adat* rules facilitate forest preservation by, for example, banning cultivation on lands around springs (*ulu ayek* or *ulu tulung*) and on areas within 100 metres of a stream bed. These rules are traditionally accepted as regulations from the ancestors.

The Dutch colonial administration in South Sumatra originally affirmed the traditional autonomous power of *margas* over their *adat* lands, but then it began to systematically take control of these native lands and forests. In 1874, the Dutch declared a *Domein-Verklaring* to claim all non-certified[14] land as state lands, and in 1916 they issued the *Bosch Ordonantie*, or forest regulation, granting the government authority to establish *bosch reserves* (reserved forests). In order to directly supervise forest lands in South Sumatra, the Dutch established the *Bosch Wezen*, or forestry office (known popularly by its abbreviation BW) in the capital city of Palembang. The BW officers registered all existing forests within the region, including the *marga*-forests, and posted BW signs along their boundaries. Three-metre zones surrounding these forest boundaries were periodically weeded to allow easy recognition of the protected areas. Armed BW escorts closely guarded the forests against violation, and patrolled the forest area borders every two months.

Marga rulers served as subordinates to the Dutch administrators in their efforts to preserve BW forests. Their routine activities included volunteering *marga* members to work under the supervision of BW officers to perform duties such as weeding the forest boundary. *Marga* rulers were also responsible for informing the Dutch administrators about any illegal occupation within the BW forests; any *pasirah* (*marga* leader) who did not carry out this duty could be punished, and a *pasirah* involved in any illegal activities within the BW forest would be dismissed from his position. Given these strict conditions, the former *pasirahs* that were interviewed are convinced that deforestation did not begin during the era of the Dutch administration.[15] Instead, the former *pasirahs* argue that significant deforestation began only in 1942, when the Japanese military occupied South Sumatra during World War II. Due to the need for supplies to support their war efforts, the Japanese military forced the rural peasants to cultivate forest lands, and *marga* rulers were unable to fend off the stream of forest cultivators who then entered these lands.

The forest lands 'borrowed' for this extra agricultural activity during the Japanese military occupation were less than 1 per cent of the total forest area at that time, but the impact was substantial. Rural farmers learned to look to forest lands as alternate farming parcels, and the influx of farmers into the forest areas has never since been successfully banned. This was the situation inherited by the Republic of Indonesia at independence in 1945. The Indonesian government initially continued the *marga* system and adopted the former Dutch government's BW forestry policy and regulations. At the beginning of the independence era, forest preservation efforts served as an expression of *marga* rulers' efforts to reestablish their traditional authority. This work did not completely restore the deforested areas, but to some extent it helped to minimize the growth of cultivated areas, especially by preventing their use by non-*marga* members who were in-migrating to South Sumatra from other parts of the country.

The Republic of Indonesia has been strongly inclined to centralize its national power, and regional autonomy has gradually decreased culminating with an executive order that abolished *margas* and established *desas* (the villages system) in their place. Several administrative problems were created by the replacement of *margas* with *desas*. For example, *desas* were established from *dusuns*, which were sub-units of the *margas* that did not have fixed borders, so the territorial boundaries of the *desas* were uncertain. In addition, administrative problems arose due to the 'power-gap' that occurred when the new *desa* rulers could not fully take over the previous duties of the *marga* rulers, particularly with respect to forest preservation, due to their lack of experience in self-governance. As a result, the new *desa* rulers found it difficult to carry out their administrative duties, and the situation worsened as the villagers became aware of the weaknesses of *desa* leadership. The weak performance of the new *desa* rulers was consistent with the expectations of rural South Sumatrans, because *marga* leadership had been based on charisma, while *desa* leaders were appointed administratively.

At the same time that the *desas* were established in South Sumatra, the new TGHK national forestry policy was imposed, replacing the old BW forest classification system with one that evaluated and reclassified national protected forest areas based on their purpose.

As a result, within the Lahat regency the forest area expanded from 165 905 ha of BW forest to 290 600 ha of TGHK forest, a 75 per cent increase. The new, expanded TGHK-forest areas included some farming areas that were previously outside of the old BW-forest areas. The immediate consequence of this expansion was an increased number of illegal farmers and an extension of illegal farming areas. This situation created additional work for government officers, who were responsible for preventing potential forest cultivators from entering a larger protected forest area, and for evicting farmers who were already settled on the lands within the new TGHK-forest boundaries.

In practice, this ambitious forest expansion effort has not been supported with adequate means to properly implement it, primarily because of tight forestry budgets. For example, the government had no funding to install new forest border signs, so the TGHK forests are visually indistinguishable from the surrounding unprotected areas. Even where the new TGHK-forest border signs have been posted, Lahat regency officials still face uncertainty in interpreting the fixed TGHK-forest boundaries. This situation reflects the lack of coordination in preserving the regency's forest areas between forestry officers, who place the signs, and Lahat regency officials, who administer the regency.

The government is also constrained by limits on the number of forestry personnel. Every forester must supervise three sub-districts. They are unarmed (except for a few special forest police), have no vehicles, and generally lack facilities to properly conduct their risky jobs.[16] Under these circumstances corruption flourishes, undermining the government's ability even to preserve the area of the former BW forests. Local government officers reportedly protect forest cultivators from being identified as illegal farmers, change forest boundary markers, and even issue fake documents of land ownership.[17] In this corrupt environment, even honest officers face internal obstacles and are powerless to change the situation. Moreover, many forestry officers and local government administrators are often indifferent about properly performing their jobs.[18]

Several military operations have been conducted in an effort to force illegal farmers to move outside the forest boundaries, but the government cannot afford to carry out such operations on a regular basis. Therefore, these occasional eviction forces only frighten cultivators temporarily, and illegal forest farming is gradually resumed once the patrols return to their normally insufficient levels.[19]

In addition, relocating forest farmers to non-protected areas has also been a difficult task for the government, since sufficiently large relocation areas are rarely available.

DEMOGRAPHIC CHANGE IN THE REGION

As of 1990, Indonesia's population was 180.5 million.[20] The population of Lahat regency is nearly 600 000, having grown by 93 per cent in the last 30 years.[21] Since the rate of natural growth is about 2.4 per cent per year, this high population increase indicates in-migration. Indeed, South Sumatra province has been a major destination site for the central government's resettlement programme. Romsan estimates that between 1934 and 1988, 741 425 individuals were relocated from densely populated areas (mostly from Java and Bali islands) to South Sumatra, and between 1980 and 1987, 296 775 transmigrants were relocated within South Sumatra, of whom 11 per cent (31 928 individuals) were resettled in Lahat regency.[22] These in-migrants make up about 5.3 per cent of the total Lahat population.[23] Unfortunately, statistics are not available for migrants who in-migrated on their own.

There is a widely-held view that transmigrants from ill-fated transmigration projects are largely responsible for the loss of forest through slash-and-burn agricultural practices.[24] However, based on information from Lahatan officials, it appears that the forest cultivators in the regency's protected areas are not from failed transmigration projects, and field observations also demonstrate that no Balinese and only a small number of Javanese are found within the study area. Moreover, the Javanese involved in forest farming within the observation area were hired by local farmers, either as labourers or via share-cropping arrangements.

SOCIOCULTURAL FACTORS CONTRIBUTING TO DEFORESTATION

Traditionally, the population of Lahat is comprised of four main ethnic groups: Lematang (or Gumai), Kikim, Pasemah and Lintang peoples. These ethnic groups speak similar languages and share the same mythology.[25] The forest cultivators in the area have more diverse origins; they include Semendo people from the adjacent

Table 6.3 Forest cultivators within the area observed
(by ethnic group)

Subdistrict	Area of deforestation (ha)	Number of households	Number of individuals	Origin	%
Pulau Pinang	257	185	744	Tanjung Sakti	50
				Pagar Alam	15
				Jarai	10
				Java	5
				Locals/others	20
Pagar Alam	560	273	1142	Manna	90
				Semendo	10
Jarai	722	518	1782	Tanjung Sakti	90
				Java	10
Kota Agung	295	191	1034	Semendo	40
				Pagar Alam	20
				Java	10
				Locals/others	30
Total	1834	1167	4720		

Source: Fieldnotes, 1991.

regency of *Kabupaten* Muara Enim, Manna people from southern Bengkulu province adjacent to South Sumatra, and people from Java island (see Table 6.3).

Tanjung Saktian people, a subgroup of the Pasemah ethnic group, are the dominant ethnic group among the forest cultivators. They are well known among Lahatans as hard-working farmers and people who share the dream of *pukul agung* (having a massive harvest), a dream that is probably derived from the concept of *jadi orang* (to 'be someone'), which can mean owning a good house, a *toko* (grocery store) at the *pasar* (market area), or a mini-truck or commuter taxi, as well as sending children to school (and perhaps to college in town, or even in Java), and, ultimately, going on a pilgrimage to Mecca to gain the religious title of *haji*. In comparison to the simple lifestyle of rural Lahatan farmers in the past, these 'materialistic desires' of the younger generation of Lahatans indicate the tremendous sociocultural changes that have occurred as the country has modernized.

In order to afford these materialistic desires, and to fulfill the goal of 'being someone', farmers plant coffee. After marriage they take their families to live in the forest, where they are isolated for years in hopes of achieving the dream of having a large coffee harvest. When they have enough money they return to their own village to build a house – a primary indicator of one's prestige. This is also the time when they begin to send their children to school within the village.

Semendo farmers have been leaving their traditional home in Muara Enim regency because of the scarcity of available farm land there, caused not only by land and population pressures, but also by the traditional *tunggu tubang* inheritance system, whereby the oldest daughter in a family remains in the village and maintains her family's properties after inheriting them from her parents, a transfer that begins on her wedding day. Once this transfer begins, all other family members still living in the household are essentially 'interlopers', and are expected to leave the house and find other resources. Many young Semendo couples are therefore forced to search for new lands outside their village, sub-district or regency, and often even outside of the province of South Sumatra. What land they do find is usually forest land, since this is the only unused land remaining.

The Javanese people found in the region include those who have been hired as farm labourers by Lahatan tenant farmers, as well as self-motivated and self-described 'adventurers' who finance their own migration to Lahat with the hope of repeating the 'success stories' of other Lahatan-Javanese coffee farmers. However, the Javanese are known for avoiding risky jobs, and since the government identifies forest cultivators as 'illegal' farmers, many Javanese shift to off-farm work in grocery stores or operating jeeps to transport agricultural products (but they may also earn more money from these new businesses than from farming). The only Javanese found working on farms within the observation area were therefore labourers or share-croppers hired by or working on behalf of Lahatan tenant farmers.

All of the forest farmers interviewed are likely to derive some sense of security from the fact that they usually begin their farming activities by asking for permission from any villagers who claim responsibility for the plots chosen. This is especially true for the Javanese, Manna and Semendo farmers. Sometimes permission comes from village or sub-district officers (always involving some kind of small compensation). To the extent that these forest farmers do have some guilty feeling about cultivating forest lands, they are not based on an awareness of environmental problems, but simply on the government prohibition of forest cultivation. Concepts like ecological equilibrium and environmental preservation are still beyond the understanding of these farmers; surviving and maintaining a livelihood are much more concrete and immediate concerns for them.

THE EFFECTS OF COFFEE PRODUCTION

The practice of monoculture production of coffee, a lucrative cash crop, by Lahatan farmers indicates the strong market incentives operating within the regency. The regency is the source of nearly 60 per cent of production in South Sumatra, which is the country's leading producer, accounting for about 25 per cent of national production.[26] After paddy rice, the domestic staple crop, coffee production is the most important crop in Lahat with respect to both the tonnage produced and the cultivated area (nearly 19 per cent of the area of Lahat). Nearly 69 per cent of the 121 000 households in Lahat cultivate coffee.[27]

Farmers indicate that the coffee market provides strong incentives for people to grow coffee, which is profitable despite the fact that international coffee prices are among the most unstable of the major agricultural commodities.[28] While coffee farmers may experience occasional losses due to price fluctuations, they may also realize major gains and typically earn enviable incomes by local standards. They also store dried coffee beans in sacks in their houses when prices are low to sell when the price rises.

Another incentive for coffee production is the limited opportunities for profitably producing other crops. Although farmers have produced good crops such as vanilla and vegetables (especially cabbage), they often do not earn a profit, especially due to transport costs. The vanilla market is distant, and vegetable producers face spoilage problems if they cannot find cheap and dependable transport. As a result farmers tend to stick to paddy rice and coffee production, and some villagers have even converted their paddy rice plots to coffee.

In response to these incentives, Lahat's farmers have increasingly cultivated highland forest land, in part because of their belief that unspoiled forest land will guarantee good harvests, whereas the soils on village lands are thought to be unfavourable for coffee production. The mountainous topography also allows cultivators to hide their illegal practices, and village lands are in any case too limited to fulfill the demand for large plots for coffee production. What little unoccupied or uncultivated land there is within the villages is often unavailable for social reasons, since 'surplus' lands are rarely sold, especially to people other than the owners' family members or neighbours. The shortage of farm plots in these villages has also increased the price of land beyond what common rural farmers can afford. As a result of all of these factors, rural coffee farmers have increasingly turned towards cultivation of forest lands where the lack of guards and the relatively uncontrolled access has allowed them to develop alternate farm plots.

Lahatan forest farmers can be differentiated according to whether they are still primarily subsistence farmers, or have entered an expansionist phase. Subsistence farmers enter the forest land as they marry, and they live within the forest, at least for the first years of cultivation, in minimal conditions. They begin by planting *padi gogo* (non-irrigated upland paddy rice) and vegetables for their daily consumption, but as soon as possible they plant coffee trees as their primary crop on an average of 2 ha of forest land. They begin to have a marketable coffee harvest sometime between three and seven years after the coffee is planted. These subsistence farmers may begin to expand their farms once they have accumulated enough capital.

Farmers will also extend their coffee farms in an effort to make up for the low prices they may receive for their crop because of low quality post-harvest processing. To protect themselves from robbery and losses due to the falling and rotting of mature beans, most farmers harvest the coffee while it is still immature. They also usually dry their coffee beans on the roads, which mixes them with dirt and gravel. As a result of this improper treatment they may receive a low price for their coffee.

MODELLING LAHAT'S DEFORESTATION

The factors influencing the dynamic relationship between Lahatan farmers and deforestation of the regency's protected areas can be grouped into push and pull factors. The push factors are those that force rural farmers to leave their villages to farm in the forest, while the pull factors are those that entice rural farmers to enter

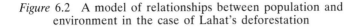

Figure 6.2 A model of relationships between population and environment in the case of Lahat's deforestation

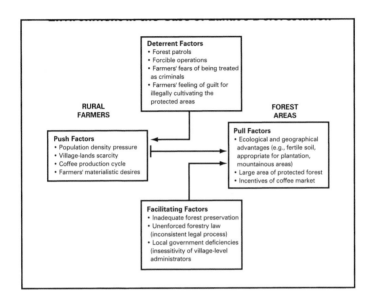

and cultivate the forest lands. The government's activities with respect to preserving the forest areas can also be grouped into deterrent and facilitating factors. A model of the dynamic interactions among the factors influencing Lahat's deforestation is shown in Figure 6.2.

A classical Malthusian approach emphasizes the push of population growth that interacts with the pull of the coffee market and the forest coffee production cycle. However, in practice it is difficult to causally separate the effects of population pressure from those of market incentives. It has, in fact, been demonstrated in several other cases that market centres develop in the areas of highest population density, and that agricultural production intensifies in those areas that are simultaneously near market centres and that support the densest populations,[29] and this pattern is also apparent in Lahat. Furthermore, it is also apparent that rural farmers' economic activities closely interact with social characteristics and the physical environment, and farmers adapt to changing population and market forces. The combination of existing push and pull factors suggests, therefore, that the failure to adequately protect forest areas is not the direct cause of deforestation, but rather a facilitating factor, and that activities to prevent further deforestation must not only be concerned with protecting the forest areas, but with development of the rural areas as a whole, so that the rural people in the region have other alternatives besides forest cultivation for sustaining themselves.

Even so, the effect of the government's administrative deficiencies on deforestation in Lahat is also significant. The decision to abolish *margas* and replace them with *desas* critically weakened the government's ability to protect the region's natural resources. As a result, village territories, including the forest areas, have been left beyond the effective control of the regency, the province, and even the central government. The appearance of such a 'power-gap' due to the replacement of older, established traditional institutions with newer, modern state agencies is not an unfamiliar factor affecting the destruction of natural resources as states seek to modernize. But this practice has frequently left natural resources unprotected at a time when traditional community norms and practices for protecting them have been weakened or abolished. In addition, the new forestry law has failed to gain community acceptance. The causes of deforestation in Lahat must therefore be seen as a complex interaction of push and pull factors, in combination with weak government capacity for protecting these resources and a failure to provide access to other alternatives for securing a livelihood.

Notes

1. This study of Lahat's deforestation is part of a project called Population–Environment Dynamics initiated by faculty from the University of Michigan School of Natural Resources and Environment, and from the School of Public Health (Department of Population Planning and International Health), in research collaboration with the Sriwijaya University Population Research Center. The project involved conducting a series of field observations in South Sumatra Province.
2. World Bank, *Indonesia Forest, Land and Water: Issues in Sustainable Development* (Washington, DC: World Bank, 1989), 157.
3. World Resource Institute (WRI), *World Resources 1992–1993* (New York and Oxford: Oxford University Press, 1992).
4. WRI, *World Resources 1990–1991* (New York and Oxford: Oxford University Press, 1990), 102.
5. WRI, *World Resources 1990–1991*; and R.C. Repetto, *The Forest for the Trees? Government Policies and the Misuse of Forest Resources* (Washington, DC: WRI, 1988).

6. P.C. West and S.R. Brechin, *Resident Peoples and Natural Parks: Social Dilemmas and Strategies in International Conservation* (Tucson: University of Arizona Press, 1991); and S.R. Brechin, S.C. Surapaty, L. Heydir and E. Roflin, 'Protected Area Deforestation in South Sumatra, Indonesia', in *Population-Environment Dynamics: Ideas and Observations*, eds G.D. Ness, W.D. Drake and S.R. Brechin (Ann Arbor, Michigan: University of Michigan Press, 1993), 226.

7. R.C. Repetto and M. Gillis (eds), *Public Policies and the Misuse of Forest Resources* (Washington, DC: WRI, 1988).

8. World Bank, *Indonesia*, note 2 above.

9. The TGHK classifies the national forests into five categories: (1) *hutan suaka alam* or conservation forests, which are set aside for conservation and include wildlife reserves, national parks and tourism forests; (2) *hutan lindung* or protection forests, which are intended primarily for watershed protection; (3) *hutan produksi terbatas* or limited production forests; (4) *hutan produksi tetap* or regular production forests, in which selective cutting is allowed; and (5) *hutan konversi* or conversion forests, which can be converted to agricultural or other uses. The first two categories are called 'permanent forests', which are designated for preservation. The previous criteria for identifying protected forest areas were elevations over 700 meters and a slope of 45 degrees or more. The new criteria employed in the TGHK system take slope, soil type and rainfall into account by calculating an indice weighted as follows:

$$\text{Slope} \times 20 + \text{Soil} \times 15 + \text{Rainfall} \times 10$$

Forests that score greater than 175 are classified as protection forests; limited production forests score between 124 and 174; and regular production forests are those that score less than 124.

10. Departemen Kehutanan, *Statistik Kehutanan Indonesia* (Jakarta: Departemen Kehutanan Republik Indonesia, 1985/1986), 118.

11. Lahat Bappeda, *Lahat Dalam Angka* (Lahat: Pemerintah Kabupaten Daerah Tingkat II Lahat, 1989), 1–2.

12. Departemen Pendidikan dan Kebudayaan, *Dampak Modernisasi Terhadap Hubungan Kekerabatan Daerah Sumatra Selatan* (Palembang: Proyek Inventarisasi dan Dokumentasi Kebudayaan Daerah, 1986/87), 33, 36.

13. Brechin *et al.*, 'Deforestation' (note 6 above); Departemen Pendidikan dan Kebudayaan, *Sistem Gotong-Royong Masyarakat Pedesaan Daerah Sumatra Selatan* (Palembang: Proyek Inventarisasi dan Dokumantasi Kebudayaan Daerah, 1981/82); W. Donner, *Land Use and Environment in Indonesia* (Honolulu: University of Hawaii, 1987); M.T. Naning *et al.*, *Evaluasi Dampak Pengembangan Perkebunan Rakyat Terhadap Potensi Sumberdaya Air di Kecamatan Jarai, Kabupaten Lahat, Indonesia* (Palembang: Pusat Penelitian Universitas Sriwijaya, 1988); *Tempo*, 15 September and 15 December 1990.

14. Written land titles are a Western legal concept that is not shared by the native population of Indonesia. Rather, the *hukum adat* or right to land is based on actual occupation.

15. Other factors that contributed to the preservation of BW forests included

low population density, the very simple lifestyle of the old-time Lahatan farmers, and the commitment of rural people to abide by the law.

16. Brechin *et al.* (note 6 above), 242.
17. One forest cultivator interviewed indicated that he has paid the amount of six sacks (600 kilograms) of coffee beans to occupy his farming plot for the last seven years.
18. *Sriwijaya Post*, 21, 22 August and 20 November 1990; and *Tempo*, 13 April and 11 May 1991, and 5 December 1992.
19. Based on *Peraturan Pemerintah* (the Government Regulation) no. 28 of 1985, articles 9 and 10, one who illegally cultivates, cuts or burns trees within the protected forest lands is liable to imprisonment for ten years, or to a fine of 100 million rupiahs (US$50 000). *Sriwijaya Post*, 1990; and *Tempo*, 1990.
20. WRI, *World Resources 1990–1991, op. cit.*
21. Brechin *et al.* (op. cit.), 233.
22. A. Romsan, *The Future Role of Public Participation in Environment and Decision Making Process: A Case Study of Indonesian Transmigration Sites in the Province of South Sumatra* (Nova Scotia, Canada: MA thesis for Dalhousie University, 1989), 54.
23. Departemen Pendidikan, *Sistem Gotong-Royong*; and Kantor Statisik Propinsi Sumatra Selatan, 'Data Statistik Propinsi Sumatra Selatan', in *Sumatra Seletan dalam Angka* (Palembang: Pemerintah Propinsi Daerah Tingkat I Sumatra Selatan, 1990).
24. C. Secrett, 'The Environment Impact of Transmigration', *The Ecologist* (1986); A.J. Whitten, 'Indonesia's Transmigration Program and its Role in the Loss of Tropical Rain Forests', *Conservation Biology* (October 1987); and WRI, *World Resources 1992–1993*.
25. Departemen Pendidikan, *Sistem Gotong-Royong*; and Departemen Pendidikan, *Dampak Modernisasi*.
26. Asosiasi Ekspotir Kopi Indonesia (AEKI), '1990 Data', in *Study of Population and Environment Dynamics in Kabupaten Lahat, South Sumatra, Indonesia*, eds S.C. Surapaty *et al.* (Palembang: Sriwijaya University Center of Population Studies, 1991); and L. Heydir, M.Y. Gani and A. Candrawati, *Identifikasi Berbagai Aspek Perladangan Liar di Kawasan Hutan di Kabupaten Lahat, Sumatra Selatan, Indonesia* (Laporan Penelitian, Palembang-Lexington-Ann Arbor: Pusat Studi Kependudukan Universitas Sriwijaya and the University of Michigan School of Natural Resources, 1990).
27. Dinas Pertanian Kabupaten Lahat, 'Data Pertanian Kabupaten Lahat', in Bappeda Lahat, *Lahat Dalam Angka*; and Kantor Statistik Kabupaten Lahat, 'Data Statistik Kabupaten Lahat', in Bappeda Lahat, *Lahat Dalam Angka*.
28. World Bank, *World Development Report* (New York: Oxford University Press, 1986).
29. C.A. Smith, 'Production in Western Guatemala: A Test of Von Thunen and Boserup', in *Formal Methods in Economic Anthropology*, ed. S. Plattner (Washington, DC: American Anthropological Association., 1975), 33.

References

Brechin, S.R., S.C. Surapaty and L. Heydir (1990) *Population–Environment Dynamics and the Power Gap Theory: Resident Peoples–Protected Area Conflicts in South Sumatra, Indonesia* (Ann Arbor: PED Symposium, University of Michigan).

Bulek, A.B. (1991) '1990 Census', in *Data Kependudukan di Propinsi Sumatra Selatan* (Palembang: Pusat Studi Kependudukan, Universitas Sriwijaya).

Departemen Kehutanan (1986/87) *Project for Development of Forestry Data and Information Systems* (Jakarta: Departemen Kehutanan Republik Indonesia).

Dove, M.R. (1985) *Swidden Agriculture in Indonesia: The Subsistence Strategies of the Kalimantan Kantu* (West Germany: Mouton).

Dutch East Indies (1933–36) *Volkstelling 1930* (Batavia: Landsdrukkerij).

Godoy, R. and C. Bennett (1988) 'Diversification among Coffee Smallholders in the Highlands of South Sumatra, Indonesia.' *Human Ecology* 16.

Lekkerkerker, G. (1916) *Land en Volk van Sumatra* (Leiden: E.J. Brill).

Russell, W.W. (1988) 'Population, Swidden Farming and the Tropical Environment', *Population and Environment: A Journal of Interdisciplinary Studies* 10.

Soewardi, B. (1983) 'Indonesia', in *Swidden Cultivation in Asia*, Vol. 1 (Bangkok: UNESCO Regional Office for Education in Asia and the Pacific).

Soewardi, B. (1983) 'Part Two: Indonesia', in *Swidden Cultivation in Asia*, Vol. 2 (Bangkok: UNESCO Regional Office for Education in Asia and the Pacific).

Surapaty, S.C., E. Roflin and L. Heydir (1991) *Study of Population–Environment Dynamics in Kabupaten Lahat, South Sumatra, Indonesia* (Palembang: Sriwijaya University Center of Population Studies).

Vatikiotis, M. (1989) 'Tug-of-War Over Trees: Indonesia Juggles Timber, Money and Conservation', *Far Eastern Economic Review* (January).

Wuisman, J.J.J.M. (1986) 'Social Change in Bengkulu: A Socio-Cultural Analysis', *Prisma: The Indonesian Indicator* 41 (September).

7 Water, Food and Population[1]

Sandra L. Postel

Unique among strategic resources and commodities, fresh water has no substitutes for most of its uses. It is essential to food production, a key ingredient in most industrial operations, and a prerequisite for human health and for life itself. Fresh water systems also provide vital ecological services that, while often hidden and easy to take for granted, are worth hundreds of billions of dollars – services that include flood protection, water purification, habitat maintenance and sustenance of fisheries.[2]

A deepening scarcity of fresh water is now a major impediment to food production, ecosystem health, social stability and even peace among nations. Each year, millions of tons of grain are grown by depleting groundwater, a clear case of robbing the future to pay for the present. Competition for water is increasing – between cities and farms, between neighbouring states and provinces, and between nations – as demands bump up against the limits of a finite supply. And valuable ecosystem functions are being destroyed by excessive damming, diverting and pollution of waterways.

With the best dam sites already developed and many rivers and groundwater reserves already overtapped, opportunities to alleviate water shortages by exploiting new water sources are limited. Sustaining food production, which is the most water-intensive of human activities and accounts for the lion's share of world water use, may be especially difficult as water scarcity spreads. With population projected to expand by 2.6 billion people over the next 30 years, and with diets in many developing countries moving up the food chain, the demand for grain – and the water to produce it – is increasing. Where this water is to come from on a sustainable basis is not obvious. It is a question few political leaders are even asking, much less rigorously attempting to answer.

WATER SUPPLY FUNDAMENTALS

The photographs of earth taken from space show a strikingly blue planet, a world of great water wealth. As it turns out, however, only 2.5 per cent of the earth's water is fresh, and two-thirds of this fresh water is locked in glaciers and ice caps. The annual renewable freshwater supply on land – that made available year after year by the solar-powered hydrologic cycle in the form of precipitation – totals some 110 000 cubic kilometers (km³), a mere 0.008 per cent of all of the water on earth.[3]

Fortunately this is still a large amount of water, but it is critical to understand where it is, what happens to it, how much of it humans already use, and how much we are likely to need in the future relative to what is available to us. Each year, nearly two-thirds of the earth's renewable supply returns to the atmosphere through evaporation or transpiration, the uptake and release of moisture by plants. This evapotranspiration (it is difficult to distinguish the two processes over large areas, hence the joint term) represents the water supply for forests, grasslands, rain-fed croplands and all other non-irrigated vegetation.

The remainder – just over one-third of the renewable supply, or about 40 000 cubic kilometers per year – is run-off, the flow of fresh water from land to sea through rivers, streams and underground aquifers. Run-off is the source for all human diversions or withdrawals of water – what is typically called water 'demand' or water 'use'. It is the supply for irrigated agriculture, industry and households, as well as for a wide variety of 'instream' water services, including the maintenance of freshwater fisheries, navigation, the dilution of pollutants and the generation of hydroelectric power. Rivers also carry nutrients from the land to the seas, and in this way help support the highly productive fisheries of coastal bays and estuaries. Thus, by virtue of the hydrologic cycle, the oceans water the continents, and the continents nourish the oceans.[4]

Although fresh water is renewable, it is also finite: the land receives roughly the same amount of water today as when the first civilizations emerged thousands of years ago. This means that renewable supplies per person decrease as population increases. Since 1950, for example, the quantity of renewable water per person has dropped by more than half because world population has more than doubled in this time. Moreover, nature delivers its water bounty unevenly, both in time and space. Asia, for example, has 36 per

Food, Water and Population

Table 7.1   Run-off and population by continent

	Total annual run-off (cubic kilometers)	Share of global run-off (per cent)	Share of global population (percent)
Europe	3 240	8	13
Asia	14 550	36	60
Africa	4 320	11	13
N. and C. America	6 200	15	8
S. America	10 420	26	6
Australia & Oceania	1 970	5	< 1
Total	40 700	101[1]	−100[1]

[1] Does not add to 100 because of rounding.

Source: Sandra L. Postel, Gretchen C. Daily and Paul R. Ehrlich, 'Human Appropriation of Renewable Freshwater', *Science*, 9 February 1996.

cent of global run-off but 60 per cent of the world's people, while South America supports only 6 per cent of the world's population with 26 per cent of the world's run-off (see Table 7.1). The Amazon River alone carries 15 per cent of the earth's run-off, but is accessible to only 0.4 per cent of world population. Fifty-five rivers in northern North America, Europe and Asia, with combined annual flows equal to about 5 per cent of global run-off, are so remote that they have no dams on them at all.[5]

According to an analysis by this author and Gretchen C. Daily and Paul R. Ehrlich of Stanford University, the total volume of run-off within reach geographically totals some 32 900 cubic kilometers. However, about three-fourths of this is flood water and is not controllable and accessible on demand, a feature that is essential for water to be useful to farmers, industries and households. To capture some of this water, engineers have built large dams and reservoirs which collectively can hold about 14 per cent of annual run-off, enough to increase the stable supply provided by underground aquifers and year-round river flows by nearly one-third. This results in a total stable and geographically accessible renewable supply – that actually available for irrigation, industries and households – of 12 500 cubic kilometers per year.[6]

Globally, water use roughly tripled between 1950 and 1990, and now stands at an estimated 4430 cubic kilometers – 35 per cent of the accessible supply (see Table 7.2). At least an additional 20 per

Table 7.2 Estimated global water demand, total and by sector,
ca. 1990

Sector	Estimated demand (cubic gallons per year)	Share of total (per cent)
Agriculture[1]	2 880	65
Industry[2]	975	22
Municipalities[2]	300	7
Reservoir Losses[3]	275	6
Total	4 430	100

[1] Assumes average applied water use of 12 000 m^3 per hectare.
[2] Estimates are from shiklomanov; see note 3.
[3] Assumes evaporation loss equal to 5 per cent of gross reservoir storage capacity.

Source: Sandra L. Postel, Gretchen C. Daily, Paul R. Ehrlich, 'Human Appropriation of Renewable Fresh Water', *Science*, 9 February 1996.

cent is used 'instream' to dilute pollution, sustain fisheries and to transport goods. So humanity is already using, directly or indirectly, more than half of the water supply that is accessible.[7]

Given that world population is projected to climb by 2.6 billion over the next 30 years – about the same number that was added between 1950 and 1990, when global water use tripled – this is a troubling finding. The health of the aquatic environment is already in a serious state of decline because of dam construction, river diversions, pollution and other human-induced changes. Even if physically and economically possible, another tripling of world water use would destroy numerous aquatic ecosystem functions, decimate fish populations and impose enormous ecological costs. Indeed, the construction of new dams has slowed markedly over the last couple of decades as the public, governments and financial backers have begun to pay more attention to their high economic, social and environmental price. Whereas nearly 1000 large dams came into operation each year from the 1950s through the mid-1970s, the number dropped to about 260 annually during the early 1990s. Even if conditions become more favourable to dam construction, it seems unlikely that new reservoirs built over the next 30 years will increase accessible run-off by more than 10 per cent.[8]

With the oceans holding more than 97 per cent of all the water on earth, desalination is often held up as the ultimate solution to the world's water problems. Yet, despite considerable growth in recent years, desalination continues to play a minor part in the global supply picture – accounting for less than two-tenths of 1 per cent of world water use.[9] Removing salt from water is highly energy-intensive, and, although costs have come down, at \$1.00–\$1.60 per cubic metre, desalination remains one of the most expensive water supply options. For the foreseeable future, sea water desalination will likely remain a lifeline technology for water-scarce, energy-rich countries, as well as island nations with no other options. But the current desalination capacity of 7.4 billion cubic metres per year would have to expand 30-fold to supply even 5 per cent of current world water use. Thus, along with towing icebergs and shipping water by tanker or in giant bags, its contribution to total water supplies worldwide is likely to remain relatively minor.[10]

WATER-STRESSED COUNTRIES

While globally constraints lie on the not-too-distant horizon, problems already exist at the country and regional levels. One broad indicator of water constraints is derived by examining a country's renewable water supply relative to its population. As a rule of thumb, countries are considered 'water stressed' when total run-off per capita drops below 1700 cubic metres per year. Exactly what the term 'water-stressed' means is a subject of debate among water analysts, but it is clear that at about this level, meeting all food, industrial and household water needs becomes difficult, if not impossible. In many, if not most, countries it is difficult to gain access to and control more than 30–50 per cent of run-off (although there are notable exceptions, such as Egypt, with the massive storage provided by the Aswan Dam.) Thus, run-off of 1700 cubic metres annually would typically translate to some 500–850 cubic metres of usable supply per capita.

On the demand side, the United Nations Food and Agriculture Organization estimates that producing the food needed for a nutritious but low-meat diet of some 2700 calories a day requires about 1600 cubic metres of water per person per year.[11] In humid climates, virtually all of this water is provided directly to the soil by natural rainfall, but in drier regions and those with distinct wet and dry seasons a portion of the needed moisture has to be supplied by irrigation water drawn from rivers, lakes or aquifers. If even a third of the 1600 cubic metres per person per year needs to be supplied by irrigation (and the actual share is far higher in dry regions), then water demands per person will be in the low end of the range of usable supply for a country with total per capita run-off of 1700 cubic metres per year.

Of course, countries have more than just food needs to meet. Estimates by Russian hydrologist Igor Shiklomanov suggest that worldwide household, municipal and industrial water use averages about 240 cubic metres per capita per year.[12] More widespread use of efficient technologies could reduce this level substantially, but the resulting savings would be partially offset by the additional water needed to meet the basic household water requirements of the more than 1 billion people for whom they are currently unmet, as well as by the higher demands arising from increasing average incomes. Assuming an average for household, municipal and industrial uses of 200 cubic metres per capita per year, and adding this to the minimum run-off required for food production in drier regions, yields a total run-off requirement of some 730 cubic metres per capita per year – in the upper range of the quantity likely to be accessible for human uses in a country with total run-off of 1700 cubic metres per capita per year.

As of 1995, 44 countries with a combined population of 733 million people had annual renewable water supplies per person below 1700 cubic metres. Just over half of these people live in Africa and the Middle East, where the populations of many countries are projected to double within 30 years (see Table 7.3). Water-short Algeria, Egypt, Libya, Morocco and Tunisia are each already importing more than one-third of their grain. With their collective population projected to grow by 87 million people over the next 30 years, their dependence on grain imports is bound to increase. Indeed, a likely scenario for much of Africa given current population projections is that more than 1.1 billion Africans will be living in water-stressed countries by 2025 – three-quarters of the continent's projected population at that time.[13]

Portions of many large countries, including China, India and the United States, would also qualify as water-stressed if water supplies and population were broken down by region. Even using national statistics, China – with 7 per cent of global run-off but 21 per cent of world population – will only narrowly miss the 1700 cubic metre per capita mark in 2030. India, the world's second most populous country, will have joined the list by then.[14] Limited water supplies combined with population growth thus appear to be eliminating the option of food self-sufficiency in more and more countries.

Table 7.3　African and Middle Eastern countries with less than 1700 cubic metres of run-off per person in 1995, with population projections to 2025[1]

Country	Internal run-off per capita, 1995 (cubic metres per year)	1995 population (millions)	Projected 2025 population (years)
Africa			
Algeria	489	28.4	47.2
Burkina Faso	1683	10.4	20.9
Burundi	563	6.4	13.5
Cape Verde	750	0.4	0.7
Djibouti	500	0.6	1.1
Egypt	29	61.9	97.9
Eritrea	800	3.5	7.0
Kenya	714	28.3	63.6
Libya	115	5.2	14.4
Mauritania	174	2.3	4.4
Morocco	1027	29.2	47.4
Niger	380	9.2	22.4
Rawanda	808	7.8	12.8
Somalia	645	9.3	21.3
South Africa	1030	43.5	70.1
Sudan	1246	28.1	58.4
Tunisia	393	8.9	13.3
Zimbabwe	1248	11.3	19.6
Middle East			
Iraq	1650	20.6	52.6
Israel	309	5.5	8.0
Jordan	249	4.1	8.3
Kuwait	0	1.5	3.6
Lebanon	1297	3.7	6.1
Oman	909	2.2	6.0
Saudi Arabia	119	18.5	48.2
Syria	517	14.7	33.5
United Arab Emirates	158	1.9	3.0
Yemen	189	13.2	34.5
Total 1995		**380.6**	**739.8**
Additional countries by 2025[2]			**618.0**
Projected total 2025			**1357.8**

[1] Run-off figures do not include water flowing in from neighbouring countries: Djibouti, Mauritania, Sudan and Iraq would have more than 1 700 cubic metres per person in 1995 and 2025 if current inflow from other countries were included.
[2] Botswana, Gambia, Ghana, Lesotho, Madagascar, Malawi, Mauritius, Nigeria, Senegal, Swaziland, Tanzania, Togo, Uganda.

Source: UN Food and Agriculture Organization, *Irrigation in Africa in Figures* (Rome: 1955); World Resources Institute, *World Resources 1994–95* (New York: Oxford University Press, 1994); Population Reference Bureau, *1995 World Population Data Sheet* (Washington, DC: 1995).

WATER FOR FOOD PRODUCTION

To the extent that water-stressed countries that are unable to grow sufficient food for their own populations can import food from other countries, they can at least partially avoid the consequences of water stress. Hillel Shuval, an environmental scientist at Hebrew University, points out, for example, that Israel has a highly successful modern economy and high per capita income even though its renewable water supply per person is less than one-fifth of the water-stress level. In part, Israel has succeeded so well with its limited supplies by importing much of its grain, which Shuval and others sometimes refer to as 'virtual water'.[15]

Indeed, with each ton of grain requiring about 1000 tons of water to produce it, importing grain becomes a key strategy for balancing water budgets.[16] Such a strategy would seem to make economic and environmental sense for countries that are short of water, since they can get much higher value from their limited supplies by devoting them to commercial and industrial enterprises, and then using the resulting income to purchase food through international markets. The Middle East, which is the most concentrated region of water scarcity in the world, imports 30 per cent of its grain.[17] As long as surplus food is produced elsewhere and nations with such surpluses are willing to trade, and if the countries in need can afford to pay for the imports, it would seem that water-short countries can have food security without needing to be food self-sufficient.

This tidy logic is shaken, however, not only by the growing number of people living in countries where water availability is a constraint to food self-sufficiency, but also by widespread signs of unsustainable water use in key food-producing regions. Groundwater overpumping and aquifer depletion are now occurring in many of the world's most important crop-producing areas, including parts of northern China where water tables are dropping one metre per year over a large area, as well as large portions of India, California and the US high plains, the Arabian Peninsula, and parts of Southeast Asia. Not only does this signal that limits to groundwater use have been exceeded in many areas, it also means that a portion of the world's current food supply is produced by using water unsustainably, and therefore cannot be counted as a reliable source over the long term. For example, in the Punjab, India's breadbasket, groundwater tables are dropping by 20 centimetres annually over two-thirds of the state, and, according to researchers at Punjab Agricultural University, 'questions are now being asked as to what extent rice cultivation should be permitted in the irrigated Indo-Gangetic Plains, and how to sustain the productivity of the region without losing the battle on the water front'.[18]

Like groundwater, many of the planet's major rivers are suffering from overexploitation. In Asia, where the majority of population growth – and thus food needs – will be centred, many rivers are completely tapped out during the drier part of the year, when irrigation is so essential. According to a 1993 World Bank study, 'many examples of basins exist throughout the Asia region where essentially no water is lost to the sea during much of the dry season'.[19] This includes most rivers in India, among them the Ganges a principal water source for densely populated and rapidly growing South Asia. China's Yellow River has gone dry in its lower reaches for an average of 70 days a year in each of the last 10 years, and in 1995 the river was dry for one-third of the year. Demand for water is exceeding the river's capacity to supply it, and crop production in the region will increasingly suffer the consequences.[20]

A third set of evidence of the limited availability of water for agriculture comes from examining global irrigation trends. Rising water development costs and the declining number of environmentally-sound sites for the construction of dams and river diversions are contributing to a worldwide slowdown in irrigation expansion. The per capita irrigated area has been steady or increasing during most of modern times, but it peaked in 1979 and has declined by about 7 per cent since then. This irrigation slowdown is a worrisome trend that is not likely to reverse anytime soon.[21]

At the same time, much irrigated land is losing productivity or coming out of production altogether as a result of salinization – that is, the steady buildup of salts in the root zone of irrigated soils. Although no firm global estimate exists, some 25 million hectares (ha) – more than 10 per cent of the world's irrigated area – appears to suffer from salt buildup serious enough to lower crop yields.[22] Salinization is estimated to be spreading at a rate of up to 2 million ha per year, offsetting a good portion of the gains achieved by irrigation expansion.[23] Indeed, David Seckler, Director General of the International Irrigation Management Institute (IIMI) in Sri Lanka, writes in a 1996 paper that 'the net growth of irrigated area in the world has probably become *negative*'.[24]

Finally, agriculture is losing some of its existing water supplies to cities as population growth and urbanization push up urban water demands. The number of urban dwellers worldwide is expected to double to five billion by 2025.[25] With political power and money concentrated in the cities, and with insufficient water to meet all demands, governments will face strong pressures to shift water out of agriculture, even as food demands are rising. Where this shift results in marginal lands or non-food crops coming out of production, or in gains in irrigation efficiency, it can be beneficial environmentally and have little impact on food security. But with competition for water increasing in many areas, more sizable shifts could reduce food production and destabilize farm communities.

The reallocation of water from farms to cities is well under way in both industrial and developing countries. In California, for instance, a 1957 state water plan projected that 8 million ha of irrigated land would ultimately be developed statewide; yet the state's irrigated area peaked in 1981 at 3.9 million ha, less than half this amount.[26] Net irrigated area fell by more than 121 000 ha during the 1980s. California officials project an additional net decline of nearly 162 000 hectares between 1990 and 2020, with most of the loss due to urbanization as the population expands from 30 million to a projected 49 million by the end of this period.[27]

In China, water supplies are being siphoned away from farmlands surrounding Beijing in order to meet that city's rising urban, industrial and tourist demands. The capital's water use now exceeds the supply capacity of its two main reservoirs, so farmers in the agricultural belt that rings the city have been cut off from their traditional sources of irrigation water.[28] With some 300 Chinese cities now experiencing water shortages, this shift is bound to become more pronounced.[29]

Growing demand in the megacities of Southeast Asia – including Bangkok, Manila and Jakarta – is already being partially met by overpumping groundwater. With limited new sources to tap, pressures to shift water out of agriculture will mount in these regions as well. Thus, the rising affluence of these and other rapidly industrializing regions is a double-edged sword, as the rising affluence and the higher consumption levels that it engenders will intensify competition for agriculture's water. In Malaysia, for instance, the number of golf courses has tripled over the last decade to more than 150, and 100 more are planned. Together, Malaysia, Thailand, Indonesia, South Korea and the Philippines boast 550 golf courses, with an additional 530 on the drawing boards. Besides chewing up farms and forests, golf courses in these countries typically require irrigation at the same time that crops do, during the dry season, when supplies are often already tight.[30]

Unfortunately, no one has tallied either the potential effects on future food production of the progressive shift of water from agriculture to cities, or the vulnerability of production arising from unsustainable forms of water use. Without such assessments, countries have no clear idea of how secure their agricultural foundations are, no ability to predict accurately their future food import requirements, and no sense of how or when to prepare for the economic and social disruption that may ensue as farmers lose their water.

BALANCING THE WATER–FOOD–POPULATION EQUATION

Preventing water scarcity from undermining future food security presents difficult challenges. A broad set of policy reforms will be needed to encourage more efficient use and allocation of water, to reduce consumption and population growth rates to sustainable levels, and to ensure that human water use and management do not undermine the functioning of critical ecosystems.

A first step is to begin pricing water more realistically. The large subsidies to water users that continue unchecked in most countries discourage efficiency investments and convey the false message that water is abundant and can be wasted, even as rivers are drying up, aquifers are being depleted and fisheries are collapsing. Farmers in water-short Tunisia pay just 5¢ per cubic metre for irrigation water – one-seventh the cost of supplying it to them.[31] Jordanian farmers pay less than 3¢ per cubic metre, a small fraction of the water's full cost.[32] Federal construction cost subsidies to irrigators in the western United States are estimated to total at least $20 billion, representing 86 per cent of total construction costs.[33] And in India, the Madras Institute of Development Studies estimates that less than 10 per cent of the total recurring costs for major and medium-sized irrigation projects built by the government as of the mid-1980s had been recovered.[34]

Although there may be sound social reasons to subsidize irrigation to some degree, especially for poor farmers, the degree of subsidization that exists today is a major barrier to achieving more sustainable water use. Charging prices that at least cover operation and maintenance costs and send a proper signal about the need for efficiency improvements is essential. There is a broad spectrum of options between full-cost pricing, which could put many farmers out of business, and a marginal cost of nearly zero to the farmer, which is a clear invitation to waste water.

One such example is the Broadview water district in California, where farmers irrigate 4000 ha of melons, tomatoes, cotton, wheat and alfalfa. In the late 1980s, when the district was faced with the need to reduce drainage into the San Joaquin River, it established a tiered water-pricing structure. The district determined the average volume of water used over the 1986–88 period, and applied a base rate of $16 per acre-foot (1.3¢ per cubic metre) to 90 per cent of this amount. Any water used above that amount was charged at a rate 2.5 times higher. In 1991, only seven out of the 47 fields in the district had any water charged at the higher level, the average amount of water applied on the district's farms had dropped by 19 per cent, and salt releases had fallen by 4000 tons. Even though farmers were still paying prices far below the water's real cost, the pricing structure nonetheless created an incentive for them to conserve.[35]

Along with improving irrigation efficiency, more appropriate pricing would help stretch the resource by promoting the re-use of urban wastewater for irrigation. This option is typically more expensive than most conservation and efficiency measures, but often less expensive than developing new water sources. Wastewater contains nitrogen and phosphorus, which can be pollutants when released to lakes and rivers, but are nutrients when applied to farmland. Moreover, unlike many other water sources, treated wastewater will be both an expanding and fairly reliable supply, since urban water use will likely double by 2025. As long as the wastewater stream is kept free of heavy metals and harmful chemicals, and is treated adequately for irrigation use with respect to disease vectors, it can become a vital new supply for agriculture. Israel, for example, re-uses 65 per cent of its domestic wastewater for crop production. Currently, treated wastewater accounts for 30 per cent of the nation's agricultural water supply, and this figure is expected to rise to 80 per cent by 2025.[36] Tunisia currently irrigates 3000 hectares with treated wastewater and plans to increase this area tenfold by the year 2000.[37] Worldwide, assuming domestic and municipal use doubles by 2025, the re-use of 65 per cent of the resulting wastewater could boost agricultural supplies by 350 billion cubic metres per year – enough, in theory, to grow 350 million tons of wheat.[38]

Along with more effective water pricing, water marketing can create incentives both to encourage efficiency and re-use, as well as to allocate water more productively. In most developing countries, water trading typically consists of spot sales or one-year lease arrangements, often between neighbouring farmers. A 1990 survey of surface canal systems in Pakistan found active water trading in 70 per cent of them. In India's western state of Gujarat, informal groundwater markets have emerged spontaneously and provide many farmers with water of high quality when needed, thus enhancing crop production. Since marketing may allow farmers who cannot afford to drill their own wells to purchase water from other irrigators, it can help provide the poor with access to irrigation water they otherwise would not have.[39] Water marketing is not appropriate or workable everywhere, however, since it requires well-defined property rights to water. And if unregulated or monopolistic, water markets can lead to overexploitation of water sources, inequalities in water distribution and exploitative prices.[40]

Along with policy reforms, stepped-up scientific research is critical, especially for raising the water productivity of the global cropbase. If the battle on the agricultural front is to be won, crop output per unit of water input will need to increase not only in irrigated farming systems, but in rainfed and water-harvesting systems as well. The actual strategies used will vary by crop, climate and the type of water-control system, but the basic aim will necessarily be the same in each: to optimize the timing and amount of moisture in the root zone and to enhance the crops' ability to use that moisture productively.

Through plant breeding and genetic manipulation, for example, scientists can hasten the process of plant adaptation to dryness. Studies have shown that if no other factors are limiting, total dry-matter production is linearly proportional to the amount of water a plant evapotranspires. Larger or deeper root systems that allow plants to take in more moisture can thus increase yield, as can adapting crops for early reproduction so as to avoid the effects of late-season droughts. New genetic techniques are making it possible to screen crop varieties for water-efficiency traits. Development of varieties with shorter growing seasons or the ability to grow in cooler periods, when evapotranspiration is lower, is also helping to improve crop water-use efficiency.[41]

IIMI and the International Rice Research Institute (IRRI), also in Sri Lanka, are two of the research centres taking up the water-efficiency challenge in a major way. A 1995 IRRI report notes that 'while the full import of the water supply problem to rice production has been recognized only relatively recently by the research community, it is now fully acknowledged there'. Future rice production will depend heavily on getting 'more rice per unit of water input'. IRRI is focusing on developing more efficient irrigation operations, technologies that reduce water consumption, and changes in the rice plant itself to improve water-use efficiency. Breeders have already shortened the maturation time for irrigated rice from 150 days to 110 days, for example, a major water-saving achievement.[42]

Better matching of crops with water of varying qualities can also enhance water-use efficiency in irrigated agriculture. In the western Negev of Israel, for example, cotton is successfully grown by irrigation with highly salty water from a local saline aquifer. The Israelis have also found that certain crops – for example, tomatoes grown for canning or pastes – may actually benefit from somewhat salty irrigation water.[43] Variation in crop tolerances raises the possibility of multiple re-use of irrigation water. In California, for example, moderately salty drainage water from a crop of average salt-tolerance is used to irrigate more highly tolerant cotton. In turn, the drainage from the cotton fields, which is highly salty, is used to irrigate halophytic (salt-loving) crops, which scientists are making considerable progress toward commercializing.[44] A variety of the oilseed Salicornia, for example, was irrigated with sea water in a coastal desert near Mexico's Sea of Cortez, and yielded seed and biomass equal to or greater than freshwater oilseed crops such as soybean and sunflower.[45]

At the international level, greater efforts are needed to assess and monitor the availability of water for food production on a worldwide basis. The time may not be far off when a global grain bank will be needed to guard against food shortfalls induced by water shortages. Particularly in Africa and Asia, water deficits will increase markedly in the coming decades. Together these regions are projected to grow by nearly 2.3 billion people by 2025; they will account for 87 per cent of projected population growth over the next 30 years. Yet many African and Asian countries are unlikely to have the financial resources to balance their water books by purchasing surplus grain on the open market, assuming such surpluses exist. Thus, just as water security is being enhanced in some regions by the establishment of water banks to cope with droughts, food security may depend on the institution of a global grain bank that can provide staple foods on concessional terms for poor, water-short countries.[46]

Finally, the conjunction of biophysical, economic and environmental limits to water use suggest that living within the limits of nature's water supply will require reduced consumption among the more wealthy social groups, and reduced family size among all groups. With nearly two out of every five tons of grain going into meat and poultry production, individual choices about diet can collectively have a major influence on how much water is needed to satisfy future food demands. And stepped-up efforts to create the conditions needed for population stabilization must be at the core of any successful strategy to achieve a sustainable and secure water future for all.

By 2025 – just a generation away – 40 per cent of the world's people may be living in countries experiencing water stress or chronic water scarcity. Successfully meeting the challenges water scarcity poses to food security will require more than incremental progress.

Notes

1. This chapter is adapted from Sandra Postel, *Dividing the Waters: Food Security, Ecosystem Health, and the New Politics of Scarcity*, Worldwatch paper no. 132 (Washington, DC: Worldwatch Institute, September 1996).
2. Sandra Postel and Stephen Carpenter, 'Freshwater Ecosystem Services', in *Nature's Services: Societal Dependence on Natural Ecosystems*, ed. Gretchen C. Daily (Washington, DC: Island Press, 1997).
3. I.A. Shiklomanov, 'World Fresh Water Resources', in *Water in Crisis: A Guide to the World's Fresh Water Resources*, ed. Peter H. Gleick (New York: Oxford University Press, 1993); renewable supply estimate from ranges in Gleick, *op. cit.*, this note.
4. See Sandra L. Postel, Gretchen C. Daily and Paul R. Ehrlich, 'Human Appropriation of Renewable Fresh Water', *Science* (9 February 1996).
5. Postel *et al.*, 'Human Appropriation', based on runoff estimates from M.I. L'Vovich *et al.*, 'Use and Transformation of Terrestrial Water Systems', in *The Earth as Transformed by Human Action*, eds B.L. Turner *et al.* (Cambridge: Cambridge University Press, 1990); Table 1 population estimates from Population Reference Bureau, *1995 World Population Data Sheet* (Washington, DC: PRB, 1995); Amazon flow from E. Czaya, *Rivers of the World* (New York: Van Nostrand Reinhold, 1981); undammed northern rivers from Mats Dynesius and Christer Nilsson, 'Fragmentation and Flow Regulation of River Systems in the Northern Third of the World', *Science* (4 November 1994).

6. Postel *et al.*, 'Human Appropriation' (note 4 above); distribution of run-off between flood flows and stable flows, and dam capacity figures from L'Vovich *et al.*, 'Use and Transformation'.
7. Postel *et al.*, 'Human Appropriation', *ibid.*
8. Number of dams from Patrick McCully, *Silenced Rivers* (London: Zed Books, 1996); potential increase in accessible runoff from Postel *et al.*, 'Human Appropriation', *op. cit.*
9. Desalination capacity in 1995 from Pat Burke, Secretary General, International Desalination Association, private communication, Topsfield, Massachusetts, 1 August 1996; growth trends from Wangnick Consulting, *1994 IDA Worldwide Desalting Plants Inventory*, Report no. 13 (Topsfield, Mass.: International Desalination Association, 1994), and Wangnick Consulting, *1990 IDA Worldwide Desalting Plants Inventory* (Englewood Cliffs, NJ: International Desalination Association, 1990).
10. Costs from World Bank, *From Scarcity to Security: Averting a Water Crisis in the Middle East and North Africa* (Washington, DC: World Bank, 1995), which gives a range of $1.00–1.50; the largest US sea water desalination plant, completed in Santa Barbara, California in 1992 and since mothballed, was estimated to produce water at $1.57 per cubic metre; energy requirements from Peter H. Gleick, 'Energy and Water', *Annual Review of Energy and Environment* 19 (1994).
11. Malin Falkenmark, 'Meeting Water Requirements of an Expanding World Population', paper prepared for Land Resources: On the Edge of the Malthusian Precipice, The Royal Society, London, 4–5 December 1996.
12. Shiklomanov, 'Fresh Water Resources', note 3 above.
13. Some of these countries benefit from rivers flowing in from neighbours, and thus are not as water-poor as their per capita internal supplies imply. Egypt, for example, has only 2.6 billion cubic metres of renewable water generated within its territory, but gets at least an additional 56 billion cubic metres more from the Nile flowing in from Sudan. These figures do not include this inflow, because when added up on a regional or global basis this would cause substantial double-counting. In general, however, countries with relatively low run-off per person also have relatively low net precipitation per person.
14. Population figures from PRB, *1995 Population*; data on water-stressed countries from Robert Engelman and Pamela LeRoy, *Sustaining Water: Population and the Future of Renewable Water Supplies* (Washington, DC: Population Action International, 1993); China's run-off from WRI, *World Resources 1994–95*.
15. Hillel Shuval, 'Sustainable Water Resources Versus Concepts of Food Security, Water Security, Water Stress for Arid Countries', paper prepared for Stockholm Environment Institute/United Nations Workshop on Freshwater Resources, New York, 18–19 May 1996.
16. Figure of 1000 tons from UN Food and Agriculture Organization, *Yield Response to Water* (Rome: FAO, 1979).
17. Tim Dyson, *Population and Food: Global Trends and Future Prospects* (London: Routledge, 1996).

18. India's Punjab from International Rice Research Institute, *Water: A Looming Crisis* (Manila: IRRI, 1995); China from Xu Zhifang, unpublished paper prepared for World Water Council – Interim Founding Committee, March 1995.
19. Harald Frederiksen, Jeremy Berkoff and William Barber, *Water Resources Management in Asia* (Washington, DC: World Bank, 1993).
20. Patrick E. Tyler, 'China's Fickle Rivers: Dry Farms, Needy Industry Bring a Water Crisis', *The New York Times* (23 May 1996).
21. Sandra Postel, 'Irrigation Expansion Slowing', in *Vital Signs 1994*, eds Lester R. Brown, Hal Kane and David Malin Roodman (New York: W.W. Norton, 1994); Gary Gardner, 'Irrigated Area Dips Slightly', in *Vital Signs 1996*, eds Lester R. Brown, Christopher Flavin and Hal Kane (New York: W.W. Norton, 1994).
22. Sandra Postel, *Last Oasis: Facing Water Scarcity* (New York: W.W. Norton, 1992).
23. Dina L. Umali, in *Irrigation-Induced Salinity* (Washington, D.C.: World Bank, 1993), cites sources suggesting that 2 to 3 million hectares a year may be coming out of production due to salinization, which, if accurate, would counteract the 2 million hectares of average annual irrigation expansion in recent years.
24. David Seckler, 'The New Era of Water Resources Management: From "Dry" to "Wet" Water Savings', Consultative Group on International Agricultural Research, Washington, DC, 1996.
25. Gershon Feder and Andrew Keck, 'Increasing Competition for Land and Water Resources: A Global Perspective', World Bank, Washington, DC, March 1995.
26. Peter H. Gleick *et al.*, *California Water 2020: A Sustainable Vision* (Oakland, California: Pacific Institute for Studies in Development, Environment and Security, 1995).
27. California Department of Water Resources, *California Water Plan Update*, Vol. 1 (Sacramento, California: Department of Water Resources, 1994).
28. Frederiksen *et al.*, *Management in Asia*; Patrick E. Tyler, 'China Lacks Water to Meet Its Mighty Thirst', *New York Times* (7 November 1993).
29. Xu Zhifang paper prepared for presentation at the first meeting of the Founding Committee of the World Water Council, Montreal, Canada, March 1995.
30. Philip Shenon, 'Fore! Golf in Asia Hits Environmental Rough', *The New York Times* (22 October 1994); see also Anne E. Platt, 'Toxic Green: The Trouble with Golf', *World Watch* (May–June 1994).
31. World Bank, *From Scarcity to Security*, note 10 above.
32. *Ibid*.
33. Richard W. Wahl, *Markets for Federal Water: Subsidies, Property Rights and the Bureau of Reclamation* (Washington, DC: Resources for the Future, 1989); see also Subcommittee on Oversight and Investigations, Committee on Natural Resources, US House of Representatives, *Taking from the Taxpayer: Public Subsidies for Natural Resource Development* (Washington, DC: 1994), which cites estimates of the total irrigation subsidy of $34–70 billion.

34. India figure from A. Vaidyanathan, 'Second India Series Revisited: Food and Agriculture', report prepared for World Resources Institute, Washington, DC, undated.
35. Described in Gleick *et al.*, *California Water*, note 26 above.
36. Shuval, 'Sustainable Water Resources', note 15 above.
37. Tunisia from World Bank, *From Scarcity to Security*, note 10 above.
38. Worldwide municipal use from Shiklomanov, 'Fresh Water Resources'; worldwide calculation assumes that 10 per cent of domestic use is consumed, leaving 90 per cent as wastewater; it also assumes that 1000 cubic metres of water are needed to produce a ton of wheat, including only direct evapotranspiration requirements.
39. Mateen Thobani, 'Tradable Property Rights to Water', FPD note no. 34 (Washington, DC: World Bank, February 1995); Tushaar Shah, *Groundwater Markets and Irrigation Development* (Bombay: Oxford University Press, 1993).
40. See Kuppannan Palanisami, 'Evolution of Agricultural and Urban Water Markets in Tamil Nadu, India', in *Tradable Water Rights: Experiences in Reforming Water Allocation Policy*, eds Mark W. Rosegrant and Renato Gazmuri Schleyer (Arlington, Virginia: Irrigation Support Project for Asia and the Near East, 1994).
41. Paul J. Kramer and John S. Boyer, *Water Relations of Plants and Soils* (San Diego, California: Academic Press, 1995).
42. IRRI, *Looming Crisis*, note 18 above.
43. Israeli examples from A. Benin, 'Utilization of Recycled, Saline and other Marginal Waters for Irrigation: Challenges and Management Issues', in *Uso del Agua en las Areas Verdes Urbanas*, eds F. Lopez-Vera, J. De Castro Morcillo and A. Lopez Lillio (Madrid, Spain, 1993).
44. Seckler, 'New Era', note 24 above; US National Research Council, *Saline Agriculture: Salt-Tolerant Plants for Developing Countries* (Washington, DC: National Academy Press, 1990).
45. Edward P. Glenn *et al.*, '*Salicornia bigelovii* Torr.: An Oilseed Halophyte for Seawater Irrigation', *Science* (1 March 1991).
46. PRB, *1995 Population*, note 5 above.

8 Population as a Scale Factor: Impacts on Environment and Development[1]

Robert Engelman

SCIENTIFIC UNCERTAINTY AND ITS POLICY IMPLICATIONS

The rapid expansion of human activities in recent decades is the decisive factor in the alteration of the natural resource base that has been available to societies for millennia. Scholars and analysts have not, however, been able to reach any agreement on the precise role that population growth plays in this alteration. They differ substantially with respect to the relative importance they place on simple human numbers, versus other aspects of human behaviour such as consumption patterns, the nature of political institutions, the distribution of wealth and technological change.[2]

This confusion stems in part from a lack of uniformity in dealing with the critical issue of *time*. Population growth during any given time period may have little impact on the use of a natural resource *within that period*. But the cumulative outcome of past growth may largely determine the scale of resource use even after a population stabilizes. The current disproportionate share of the world's energy used by the US, for example, owes more to the 3 per cent annual population growth that the country experienced during the nineteenth century than to the 1 per cent annual growth it experiences today. Of course, the average American also uses many times more commercial energy than the average resident of most developing countries. But this fact would be of only passing consequence had the US population stabilized at its 1820 level of 9.6 million.[3] By the same reasoning, the environmental impacts of today's rapid population growth in some countries may not be fully evident until after their populations have stabilized, presumably at

126

a time when their level of development has advanced and per capita consumption is higher than it is today.

For a variety of reasons – especially the complexities and uncertainties inherent in both human and material systems – a precise quantification of the role of population dynamics in natural resource use and disposal is not possible. Aiming for precision in this field is nevertheless valuable in that it improves understanding and stimulates dialogue. But many policy-makers are being persuaded of the importance of population by the realities of their everyday experiences within their own countries – growing conflicts over scarce water supplies, for example, or food production that fails to keep up with increasing demand – rather than by abstract scientific analysis and debate. The intricacies of the relationships may elude them, but they know that there is an intimate connection between the size and change of their populations and the resource problems they must manage. Not surprisingly, it is increasingly presidents, agricultural ministers, water planners and fisheries department directors who are calling attention to the unsustainability of their nations' population growth rates.

Yet even this recognition and the subsequent attempts to define the issues and needs are complicated by the difficulty of distinguishing between local and remote population–environmental linkages – for example distinguishing between the effects of population growth in San Francisco on the immediate region, and its remote impacts on, say, the rain forests of Irian Jaya. More than half of all of the world's population growth, for example, is occurring in countries with tropical forests such as Zaire, Brazil, Indonesia, the Philippines, Mexico and the nations of Central America. While some assessments assign the dominant responsibility for loss of forest cover to population growth – usually defined as an increase in the numbers of landless farmers on the perimeter of forests – others blame logging or other commercial resource extraction, including the spread of cattle ranches. Discussing the ultimate and proximate causes of the loss of forests in these countries thus illustrates the perils of overly narrow views of the nature and location of the population growth that is responsible for the changes. Properly speaking, all of these changes are linked to population growth somewhere, whether it be the local increases in inhabitants of the countries where the forests are located, or more remote changes that lead to increasing global demand for tropical woods and beef, which are clearly related to both population and economic growth worldwide.

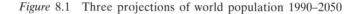

Figure 8.1   Three projections of world population 1990–2050

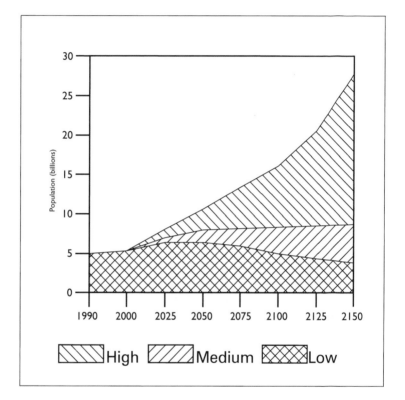

Source: United Nations, 1992, *Long-Range World Population Projections: Two Centuries of Population Growth 1950–2150* (New York: United Nations).

Based on three scenarios of future birth, death and migration rates, the UN has generated three possible projections (not predictions) of world population trends from 1950 to 2150. These projections suggest that world population in 2025 could be as low as 7.6 billion, or as high as 9 billion, with an intermediate projection of 8.3 billion.[4] In 2150, the projections suggest that human population could still be growing rapidly (high projection), roughly stable at around 12 billion (medium projection), or declining gradually (low projection) based on an assumption of low global average fertility rates (see Figure 8.1).[5] These are, of course, global figures; the impact within certain countries is far more dramatic. Based on the intermediate scenario, Ethiopia could be home to 194 million people by 2050, nearly four times its current population, and Pakistan might have 381 million people, nearly three times its present size. The impact of populations of these sizes on local resource bases – many of which already show evidence of stress and potential collapse – strains the imagination.

The point here, however, is that population projections are, in the words of Joel E. Cohen, 'conditional predictions', dependent on a number of assumptions that may or may not prove valid.[6] One such assumption is that population size can never reach a point where negative ecological and developmental feedback would lead to increases in death rates. This assumption is clearly erroneous, but because no one knows how to correct it, it remains fundamental to most population projections. For this and other reasons, population projections should be used with the understanding that they are mathematical constructs, and can only provide a general idea of, not predict with certainty, future patterns of population growth.[7]

POPULATION–ENVIRONMENT INTERACTIONS

The existing scientific literature on population–environment interactions tends to support the idea that population density does not alter environments directly or in a linear fashion. Rather, population impacts are mediated by variables such as social organization, government policies, dominant technologies and personal consumption behaviour. The connections and interactions are complex, indirect, and vary according to location and time-scale. Often the causal sequence connecting population growth with changes in the environment is rather long, particularly with respect to the impacts of population dynamics at the national level.

One useful approach to understanding population-environment interactions is to consider the changes in per capita availability of such key natural resources as renewable fresh water and arable land. Although these per capita figures are only averages and do not reflect disparities in access to resources, they constitute an important first step in understanding population–natural resource linkages. Per capita availability establishes a background level of resource abundance or scarcity against which all politics, economics and inequalities in power and wealth are played out. Moreover, per capita availability can be used as a kind of research 'control'. The concept helps analysts understand where nations would be in the absence of the many factors that, in reality, produce unequal access to resources. The per capita availability indicates what each resident would have to work with if societies ever approached the elimination of unequal access, and instead expanded *access* (the real capacity to use a resource) to the point that it approached *availability* (the theoretical capacity to use the resource).

Here again, it is useful to consider the growth of human *numbers,* not as a determinant of aggregate behaviour, but as a factor that expands the scale and extent of human activity by which real women, men and children have real impacts on the environment. With population growth, the scale of such activities and impacts tends to expand, while the availability of physical resources remains static or even declines. Being inventive, human beings typically develop new ways to use these resources more efficiently or imaginatively, but the scale imbalance – growing human activity against the backdrop of finite natural resource bases – remains. In the absence of *continuously successful* efforts to improve the efficiency of use or to develop suitable substitutes for specific natural resources, population growth tends to push the scale of human activity up to and beyond the point where collisions with natural resource limits occur.

THREE CRITICAL NATURAL RESOURCES: WATER, CROPLAND, FISH

This is anything but a hypothetical formulation, as national and local policy-makers the world over can attest. In country after country, both wealthy and poor, the supplies of renewable fresh water, cropland and fish stocks are increasingly constraining the location and extent of settlement, agriculture, industry and related development.

Because renewable fresh water – that which is provided by the hydrological cycle as rain and snow falling on land – is finite, the amount of this water available per person declines as population grows. Although renewable water can and must be used much more efficiently, it is inevitable over the long term that population growth will result in increased human withdrawals from streams and rivers. At the point that a watershed becomes 'closed' – that is, when all water is being withdrawn and 'consumed' (either through evaporation or severe pollution) – the entry of any new user requires that other users withdraw less than they did previously. This ongoing subdivision of finite water supplies can be particularly difficult – and even hazardous – in the many watersheds that cross country borders. Nine countries, for example, share the waters of the Nile River basin with Egypt, which sees the source of 97 per cent of its fresh water threatened by upstream development efforts. Until just a few decades ago, population levels were such that reasonable water development scenarios posed little threat to Egypt's security. But at the population levels projected for the Nile basin in the twenty-first century, every drop of the Nile waters will be needed by Egypt to produce its food, cool its industrial machinery and keep its citizens healthy, at the same time that increasing upstream needs may stimulate developments that will reduce this flow. Depending on the foresight of leaders in the Nile watershed, this situation could generate either an energetic diplomatic effort, or conflict.

Hydrologists use the benchmark of 1000 cubic metres of renewable water per person per year as the minimum supply level that defines water scarcity in a country; below this level it is exceedingly difficult for a developing country to make the transition to full industrialization. A country may be able to base its development on non-renewable water supplies, such as those offered by groundwater aquifers that filled in the ancient past but are no longer recharged effectively under current climatic conditions. But this is hazardous, as these aquifers eventually will become too depleted to use – perhaps at a time when both population and economic development have reached levels where the loss of dependable water supplies would threaten disaster. To ensure sustainability, societies should only depend upon renewable water.

Data for both population and total annual renewable water is available for 149 countries, and it can be used to estimate how many people will be living in countries with conditions of scarcity (fewer than 1000 cubic metres of water per person per year) or water stress (between 1000 and about 1700 cubic metres per person per year). As shown in Figure 8.2, both the number and the proportion of people living in water-short countries will depend to a large extent on which trajectory of population growth the world follows in the coming century.[8] The UN's 1994 medium population projection suggests that, by the middle of the coming century, 4.4 billion people in 58 countries will be experiencing either water scarcity or water stress, out of a projected world population of 10 billion.

Figure 8.2   Population experiencing water scarcity 1990–2050

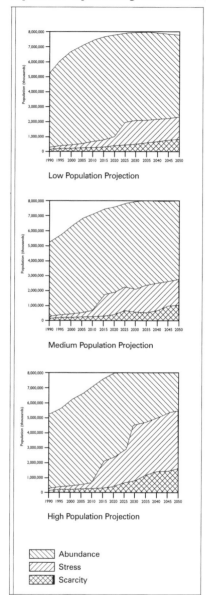

Source: Robert Engelman and Pamela LeRoy, 1995, *Sustaining Water: An Update*, Population Action International, Washington, DC.

Under the high projection, 65 per cent of the world's population would be living in 66 water-short countries (7.7 billion out of a total world population of 11.9 billion people). Under the low projection, 44 per cent of the total world population, or 3.5 billion people, would be living in 51 water-short countries.

Mexico City offers a compelling example of the impact that growing water scarcity can have on urban environmental quality and development prospects. Less than 1000 years ago, the Aztec Indians settled a verdant group of islands connecting two mountain lakes. Today the lakes have long since disappeared under pavement, and the megacity that has replaced them is home to more than 15 million people. About 70 per cent of their water is supplied by underground aquifers, which are continuously being depleted causing the city to settle and sink. At the same time this raises the risk of contamination by agricultural and industrial chemicals, which would be too expensive to clean up once it occurs. Some water is drawn over mountain passes by pipelines, but this is expensive, and few new sources of water are available as population continues to grow in the regions surrounding the city. Mexico City could recycle its water more effectively, but even so it has no long-term prospects for matching renewable water supply to the growing needs of its expanding populace.[9]

A similar relationship applies to the per capita availability of farmland, which most developing countries rely on to produce the bulk of the food consumed domestically. Unlike fresh water, food can be traded in large quantities and transported long distances. Yet any country forced to rely on external sources of food must either produce some tradable good that is at least as valuable as food, or depend on the whims of international charity for food aid. Policy-makers in most developing nations therefore prefer to achieve rough self-sufficiency in food (accounting for imports and exports).

Since the dawn of agriculture, the global cultivated area has grown roughly in proportion to human population, but in the past few decades the expansion of the cultivated area has slowed. World population now grows at eight times the pace of the total cultivated area – 1.6 per cent compared to 0.2 per cent – guaranteeing that presently-cultivated land must be worked much more intensively to bridge the gap between the two rates.[10] As the agricultural frontier closes, agricultural intensification is replacing agricultural extensification.

Without special effort, this intensification results in a depletion of soil-based carbon, which is the food on which vast and diverse populations of arthropods and micro-organisms depend. There is very little research examining the dynamics of soil biodiversity, but there is increasing evidence that a wealth of species help to cycle essential nutrients between bedrock, soil and atmosphere. As monoculture expands in farming areas, both food-related and associated biodiversity decline, often necessitating greater use of chemical pesticides. At the same time, loss of nutrients resulting not only from topsoil loss but from the simple process of transporting crop material away from the soil on which it grows requires increasing fertilizer use, which is expensive and may or may not make up for the ongoing loss of soil-based micronutrients. These micronutrients may be critical not only for yield, but for the health and disease-resistance of both the crops and the human beings who ultimately consume the grains, fruits and vegetables.

Arable land, like renewable water, is a finite natural resource, so similar calculations can be made about the future per capita availability of arable land. No society has managed to feed its population with less than 0.07 hectares (ha) of arable land per person without using the kind of intensive, chemical-based agriculture common in western Europe, Japan and the midwestern US. Based on this benchmark and the UN population scenarios, if the global supply of arable land remains constant at 1990 levels (it is not expected to grow by much, and could easily decline), between 1.6 and 5.5 billion people may be living in countries with scarcities of arable land by 2050 (see Figure 8.3).

Technological innovation in agriculture has of course boosted yields in the past, and is likely to do so in the future as well, but the past may not be an entirely reliable guide. Farmers are stressing limited cropland as never before in history, and intensification must not only continue but accelerate to keep up with population growth, economic growth and the drive to reduce malnutrition. Confidence that this task will be easy is unwarranted.

Most of the world's current population growth is occurring in countries – including India, Pakistan, China, Mexico and Ethiopia – that have almost no potential for adding to their cultivated land. Indeed, these countries are actually losing arable land to the ubiquitous spread of factories, office buildings, shops, homes and highways. The food that will be required to feed a world population of 8 billion or more in the next century will have to come almost entirely from existing farmland. Yet that farmland is not only being lost to development, it is also drying up, taking on salt and

Figure 8.3 Arable land scarcity 1990–2050

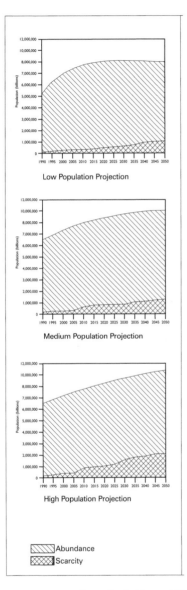

Source: Robert Engleman and Pamela LeRoy, 1995, *Conserving Land: Population and Sustainable Food Production*, Population Action International, Washington, DC.

toxic metals, or losing critical topsoil and nutrients. Between 1945 and the early 1990s, an area of land the size of China and India has lost productive potential, while 25 billion tons of topsoil wash off the land each year. With farmland, as with renewable fresh water, sound population policies are not an alternative, but only a supplement to the many steps needed to put resource use on a sustainable basis. Yet only when demand for food is roughly stable will the need to achieve food security and the need to conserve soil and water cease to be at odds to some extent.

The state of the world's fisheries provides similar evidence of the collision between growing human needs and limited natural resources. The world's oceans and rivers are unlikely to be able to supply more than 60 million metric tons per year of fish as food for human consumption. This is actually slightly above the level at which the global fish catch has stabilized since about 1989. Rapidly growing aquaculture production contributes roughly another 16 million tons today, and it is likely to contribute more in the future. But how much more? Calculations by Population Action International, based on the UN population projections and the assumption of constant global per capita fish consumption, indicate that under the medium and high population projections, by the middle of the next century fish farmers would need to produce more food from fish than all of wild catch that the world's oceans and rivers now contribute. Only under the low population projection would wild-caught fish still dominate, with aquaculture contributing 'only' 43 million tons of food (see Figure 8.4). Most fish experts are sceptical that aquaculture can go much beyond twice its current output, if that, because of limitations on and competition for suitable land for ponds and pens and a dependable water supply, and the growing challenge of keeping cultured fish in good health.[11]

The implications of this calculation for the people of developing countries are sobering. Until populations stabilize, the cost of fish is very likely to continue to rise, pricing the poor out of what historically has been among the few inexpensive sources of high-quality animal protein. Stabilizing population alone would not reverse the decline of fisheries, but as with other critical natural resources that are vulnerable to growing human demand, a combination of population and conservation strategies will be needed to allow fish to once again become a nutritious food that is available and affordable to all.

Figure 8.4   Food from fish 1995–2050

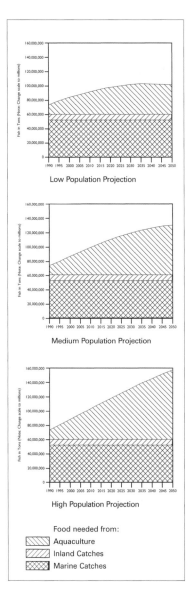

Source: Robert Engelman and Pamela LeRoy, 1995, *Catching the Limit: Population and the Decline of Fisheries* (Wallchart), Population Action International, Washington, DC.

THE CONVERGENCE OF NORTH AND SOUTH

Water, cropland and fisheries are three of the most obvious examples of critical natural resources placed at risk by national and local population growth once certain natural thresholds are approached or crossed. Although it is characteristic of both demographic and economic disciplines to separate the world's nations into two or three major groups, one cannot study these resources without being struck by the convergence of population-related resource problems in both the wealthier countries (the 'North') and those in the process of development (the 'South'). Water shortages and hazardous declines in water quality are experienced in many or most countries, and farmland suffers from soil erosion and nutrient loss and is converted to other uses nearly everywhere as well. Fisheries are declining worldwide.

The long-standing and frequently repeated idea that environmental damage worldwide is caused predominantly by the wealthiest 1 billion people in the world and the poorest 1 billion is an oversimplification of reality. All people, at all levels of incomes, have environmental impacts that vary by income level as well as by cultural, technological and even individual human differences. The dramatic increase in the middle class within developing countries, along with the onward march of globalized trade and globalized culture, are demonstrating the wide range of stresses that large populations of all income levels can have on local environments. And these factors also affect governments' capacities to manage environmental change while trying to encourage economic and social development. The immense problems that Mexico City faces with air pollution, for example, demonstrate the complexity of the linkages between population, consumption and the environment. Is this a population problem, a consumption problem, a technological problem or a policy problem? Is it a problem related to wealth, or to poverty? It is all of these and more – as are most environmental problems that policy-makers must struggle with.

RESPONSIBILITY AND THE CAPACITY TO RESPOND

As these examples of population–environment linkages indicate, once key natural thresholds are approached or crossed, further population growth tends to limit the possibilities for effective response to resource scarcity. Governments are often forced into 'lose–lose' policy choices because no available option can quickly restore a natural resource base or environmental condition to its previous state – at least not without trampling on legitimate interests. Ultimately, the effects of good policies and programmes can only be enduring – and the dilemma of irresolvable problems avoided – if population is addressed as a critical factor within a matrix of other key factors influencing the environment.

Implementing sound population policies on this basis will help determine the development prospects for all nations. The stabilization of human population will mark the dawn of an era in which problems of environmental conservation and economic development no longer pull governments in conflicting directions, or accelerate them towards disaster even as their societies are working to resolve them. In addition, population policies centred on human development and individual freedom would merit support even if they had no impact on population growth rates, because they improve well-being in multiple ways that have nothing to do with demographic change. Such policies are all the more urgent because of the close interdependence between human population dynamics and the integrity of natural resources critical to prosperity, health and our very survival.

Notes

1. This paper was presented (in a slightly different form) at the Conference on Population, Environment and Development, sponsored by the Tata Energy and Resources Institute, Washington, DC, 13–14 March 1996.
2. Pamela LeRoy, *People, Numbers, Impacts: An Annotated Bibliography of Scientific Literature on Population and the Environment* (Washington, DC: Population Action International, 1995). This bibliography contains sections listing surveys of the literature, compilations and theoretical and conceptual works.
3. US Census Bureau, data sheet on historic US population, 1994, included in personal correspondence.
4. United Nations, *World Population Prospects: The 1994 Revision* (New York: United Nations, 1994).
5. United Nations, *Long-Range World Population Projections: Two Centuries of Population Growth 1950–2150* (New York: United Nations, 1992).
6. Joel E. Cohen, *How Many People Can the Earth Support?* (New York: W.W. Norton, 1995).

7. For a fuller discussion of the limitations of population projections, see Robert Engelman, 'Population Prospects', in *Population and Global Security*, ed. Nicholas Polunin (Cambridge: Cambridge University Press, 1996).
8. Robert Engelman and Pamela LeRoy, *Sustaining Water: An Update* (Washington, DC: Population Action International, 1995). See also Robert Engelman and Pamela LeRoy, *Sustaining Water: Population and the Future of Renewable Water Supplies* (Washington, DC: Population Action International, 1993).
9. Nancy Nusser, *Des Moines Register* (28 January 1996), as summarized in *Greenwire* (6 February 1996).
10. Robert Engelman and Pamela LeRoy, *Conserving Land: Population and Sustainable Food Production* (Washington, DC: Population Action International, 1995).
11. Robert Engelman and Pamela LeRoy, *Catching the Limit: Population and the Decline of Fisheries* (wall chart) (Washington, DC: Population Action International, 1995).

Part II
The Equation
Out-of-Balance

9 Population Dynamics Revisited: Lessons for Foreign Aid and US Immigration Policy[1]

Virginia Deane Abernethy

Contending theories of population dynamics are the subject of some of the liveliest disputes in academia. The received wisdom is essentially that as stated by Shaukat Hassan:

> Population growth is an effect of underdevelopment arising from the conflict between the demand for social security and the skewed development process. Such a conflict can be resolved only by economic growth and social development.[2]

But I must regretfully differ. Does Hassan believe that rampant population growth afflicted human societies throughout history, when all were 'underdeveloped'? Nothing could be more misguided. In fact, the human societies that endured did so, in part, by developing cultural mechanisms to keep their population size in rough balance with the sustainable carrying capacity of their environment. 'Development', on the other hand, is likely to be destabilizing, leading to surges in population growth because it disturbs the average person's understanding that well-being is optimized by limiting family size. Industrialized, agricultural, hunting and nomadic societies do not differ on this dimension.

Episodes of rapid population growth occur now – just as they did in the past – because some technical, economic or political stimulus makes ordinary people believe that larger family size is affordable. That is, perceived expansion of economic opportunity – particularly if it has the quality of a windfall – raises fertility. Sometimes such economic optimism causes population growth to overshoot carrying capacity, as occurred in thirteenth-century western Europe, and again in the eighteenth and nineteenth centuries, particularly in Ireland. The overshoot is followed by economic and social collapse and restoration of cultural mechanisms that limit successful reproduction.[3] If environmental degradation is not too far advanced, the society returns to a steady state, which can include oscillations around a mean.

In light of this relation, it is apparent that development aid can have harmful effects if it leads to misperceptions about economic opportunity; assistance that sends unrealistically rosy signals about economic prospects may ultimately be seriously destructive for the very people it is meant to help. Developing a better understanding of the connection between the perceived expansion of opportunity and rising fertility can help to avoid such harmful effects.

Overpopulation afflicts most countries in the world today, but it remains primarily a local problem – a paradox that this chapter seeks to explain. The solution – reproductive restraint – is also primarily a local issue; it grows out of a sense that resources are shrinking. Under these circumstances, individuals and couples often see limitation of family size as the most likely path to success.

PROGRESS AND POPULATION

Many scholars, both ancient and modern, have known that actual family size is very closely linked to the number of children people want.[4] World Bank analyst Lant Pritchett asserts that 85 to 90 per cent of actual family size is explained by parents' family-size targets. He argues that, 'the impressive declines in fertility observed in the contemporary world are due almost entirely to the equally impressive declines in desired fertility'.[5] Cross-cultural and historic data suggest that people have usually limited their families to a size consistent with living comfortably in stable communities. If left undisturbed, traditional societies survive over long periods in balance with local resources. These societies survive in part because they maintain themselves within the carrying capacities of their environments.

However, the perception of limits imposed by the local environment is easily neutralized by the promise of prosperity. The late Georg Borgstrum, a renowned food scientist and a much decorated specialist in Third World economies, explained that a number of civilizations, including India and Indonesia, 'had a clear picture of the limitations of their villages and communities' before foreign intervention disrupted the traditional patterns. Technical-aid programmes 'made them believe that the adoption of certain technical advances [was] going to free them of this bondage and of dependence on such restrictions'.[6] Economic expansion, especially if it is broad-based and is introduced from outside of a society, encourages the belief that formerly recognized limits can be discounted, that everyone can look forward to prosperity, and, more recently, that the West can be counted upon to provide assistance to and an escape valve for excess population.

The perception of new opportunity, whether due to technological advance, expanded trade, political change, foreign aid, moving to a richer land or disappearance of competitors (who move away or die), encourages larger family size as shown by the examples listed in Table 9.1. Families eagerly fill any apparently larger niche, but the extra births and the consequent population growth often overshoot actual opportunity. Increase beyond a sustainable number is an ever-present danger, because human beings take their cues from the opportunities that are apparent today, and are easily fooled by change. Relying on what is near in space or time, we find it difficult to calculate the long-term momentum of population growth, the limits to future technological advance and the inexorable progression of resource depletion.

The appearance (and short-term reality) of expanding opportunity takes various guises. Successful independence movements and populist coups are prominent among the kinds of changes that carry the message that times are good and getting better. China launched its euphoric interlude with the expulsion of the Nationalists in 1949. Communism triumphed, and with it the philosophy that a greater nation required more people. Both the fertility rate and population size rocketed upward. The population of mainland China, estimated at 559 million in 1949, grew to 654 million by 1959, whereas in the preceding 100 years of political turmoil and war the average growth rate had been just 0.3 per cent per year.[7] Both lower mortality and higher fertility contributed to the increase. Judith Banister wrote that '[f]ertility began rising in the late 1940s, and was near or above 6 births per woman during the years 1952–57, higher fertility than had been customary' in prior decades. Banister attributes China's baby boom to the end of war and to government policy:

... the land reform of 1950–51 redistributed land to landless peasants and tenant farmers, perhaps triggering the idea that they needed more family laborers to work the land and that they would be able to feed more children from their newly acquired land.[8]

Table 9.1 Stimulants of fertility increase and decline

Country/Region	Fertility increase			Fertility decrease		
	Time period	Stimulant	Impact	Time period	Stimulant	Impact
United States	Early 1600s	Material abundance for immigrants from England and France – fluctuations follow economic cycles related to trade and agricultural reform	Fertility rates of immigrants much higher than in their home countries	1635–1815	Free land is all taken up, land prices rise, farms become smaller, labour is less valuable	Decline to less than 20 births per 1000[1]
	1947–1961	Perception of expanding economic opportunity (also in western Europe slightly later)	Baby boom	1930s	Great Depression, endemic pessimism	Low fertility in western countries
	Present	Greater employment and welfare opportunities for immigrants relative to their countries of origin, while incomes for native born are stagnant	Native-born fertility nearly constant at 1.8; immigrant rates higher than native born members of their own ethnic group; Mexican immigrants' fertility rates higher than native-born and higher than the rate (3.1) in Mexico[2]	1970s–1990s, esp. post-1973	Economic opportunity levels off, real incomes stop rising	Native-born fertility drops below replacement levels[3]
Cuba	post-1959	Fidel Castro's triumph leading to new government policies, especially redistribution of wealth, and perceptions of a more prosperous future	Fertility increase[3]	1970s and 1980s	Communism's inability to deliver prosperity	Below replacement level fertility[4]

Place	Date	Cause	Demographic response	Date	Cause/condition	Fertility outcome
Haiti	1970s and 1980s	Interruption of family-planning programmes, (billions of dollars of aid from the US in early 1980s, 1986 departure of Baby Doc, accelerating emigration which relieved competition for resources, produced cash remittances[5]	Higher rates of family formation and family stability. Fertility 5.5 increases from children per woman in the mid-1970s, to 7 in early 1986[6]	Early 1990s	Flotillas of emigrants to Florida were first accepted, but then turned back by Presidents Bush and Clinton	Fertility rates decline
	1994	US-assisted deliverance from Cedras, reinstatement of Aristide, presence of American troops	May have briefly spurred fertility	1995–present	Recognition the troops will leave; subsidies and perceptions of increasing mortality	Fertility drops to just 4.5 children per wioman[7]
Europe	500s–800s	Advances in technology (stirrups, collar harnesses, and nailed horseshoes) leading to increased agricultural productivity, improved nutrition, and economic recovery that ended the Dark Ages	Population triples in England and France (1050–1350)[8]	1980s–1990s		
Ireland	mid-1700s–mid-1800s	Introduction of potato, increased agricultural productivity, subdividing plots for sons	Earlier marriage (two-thirds marry by age 25 in 1830), baby boom[9]	1846–World War II	Land scarcity and potato famine lead to later marriage	Fertility declines; only 10% marry by age 25 in 1851[10]
Italy, Spain, Greece, Portugal				1846–World War II	Sense of scarcity or relative deprivation (poverty in comparison to one's reference group)	Depression of fertility
Eastern Europe and Russia				1980s–1990s	Economic restructuring, declining government consumption subsides, perceptions of rising infant mortality	Lower fertility, greater demand for avoiding pregnancy[12]

Table 9.1 continued

Country/Region	Fertility increase			Fertility decrease		
	Time period	Stimulant	Impact	Time period	Stimulant	Impact
East Germany				1980s–1990s	Economic shocks of reunification, end of socialism, immense unemployment	45% decline in fertility in 1991–92, total decline by 50% in 5 years to just 0.8% births per women (some recovery expected)[13]
Hungary				1980s–1990s	Decline in the standard of living	Fertility declines from 2 births per woman in 1980 to 1.6 in 1996[14]
Turkey	1950s	Land redistribution	Formerly landless peasants increase family size[15]			
China	1949–1959	Triumph of communism, end of war, need for more workers to create a great nation and take advantage of land-reform policies	Fertility rises to over 6 children per woman; Population increases from 559 million in 1949 to 654 million by 1959[16]	post-1959	Famine (30 million people in early 1960s), one-child-per-family policy (1979)	Fertility and family size decline[17]
Malaysia	post-1957	Independence democratization, Malay gain political power	Fertility rates, formerly less than others (5.5), stay high, Chinese and Indian rates drop from 7.5 to 3 births per woman by mid-1980s[18]			

India	1550–1600	End of Mongol invasions and war, new trade opportunities	Accelerating rates of population growth			
	1947–1980	Independence and development assistance from USSR, World Bank and International Monetary Fund	Even faster population growth[19]			
Africa	post–World War II	Independence, world's highest per capita levels of development assistance, improved health care, higher literacy, economic optimism	World's highest fertility rates, almost 7 children per woman in the 1950s and 1960s[20]			
Sahel	1950s and 1960s	Donor drilling of deep water wells, leading to increased herd size greater ease of obtaining bride prize	Earlier marriage, higher fertility, even faster population growth[21]			
Rwanda	1962–1990s	Independence, dispersal of people into new agricultural lands and neighbouring countries leading to a sense of new opportunities, liberation from territorial constraints	Fertility reaches 8.6 children per woman in 1987[22]	1980–present	Agriculture extensification impossible; intensification possibility limited economic decline	Higher contraceptive rates; fertility declines[22]
Ghana	1980s–1993	Emigration option remains open	Fertility remains high (6.2 births per woman in 1993)	1994–1996	Emigration option shut off	Fertility falls to 5.5 births per woman[23]
Zimbabwe	1980s	Independence, government promotes population growth	One of highest population growth rates in the world[24]	Late 1980s	International economic retrenchment, government support for family planning, high costs of supporting a family	Some men recognize value of limiting family size[25]

150   *Population Dynamics Revisited*

[1] See the following: T.T. Poleman, 'Population: Past Growth and Future Control,' *Population and Environment* 17 (1995): 29; and B.J.L. Berry, 'From Malthusian to Demographic Steady State in Cincord', *Population Development Review* 22 (1996): 207–230.

[2] V. Abernethy, *Population Politics: The Choices that Shape our Future* (New York: Plenium Press, 1993); *World Population Data Sheet*, Population Reference Bureau, 1994–96.

[3] S. Diaz-Briquets and L. Perez, *Cuba, the Demography of Revolution* (Washington, DC: Population Reference Bureau, 1981), 15.

[5] K. Freed, 'Hatians: Hopes Riding on Boats, New President', *Grenberg Tribune Review* (from *Los Angeles Times*)(26 November 1992): I.H. ooper, 'For Haiti, Corruption, Crime and Deprivation Hobble Economic Plan', *Wall Street Journal* (4 November 1992): 1.

[6] J.P. Guengant and J. May, 'Tendances de la Fecondite en Haiti', *Cahjiers Quebecois de Demographie* 21 (1992): 170

[7] V. Abernathy, 'Population Theory and Future Population Size', *Environmental Pollution*– International Conference on Environment and Population (ICEP) 3; ed. B Nath (London: European Centre for Pollution Research 1996).

[10] K. Davis, 'The Theory of Change and Response in the Modern Demographic History', *Population Index* 29 (1963): 345–64.

[11] Carl Haub, 'World Population Expected to Reach 6 Billion in early 1999', *Population Today* 24 (1996): 1–2.

[12] Cristoph Conrad, M. Lechner and W. Warner, 'East German Fertility after Unification: Crisis or Adaptation', *Population and Development Review* 22 (1996): 331–358; Haub, 1996.

[14] V. Abernethy, 'The Fickle Finger of Fate', *Population and Environment* 18 (1997).

[15] B. Aswad, 'And the Poor Get Children: Radical Perspectives on Population Dynamics', *Monthly Review Press* (1981).

[16] Judith Bannister, *China's Changing Population* (Stanford, California: Stanford University Press, 1987).

[18] P. Govindasamy and J. DaVanzo, 'Policy Impact on Fertility Differentials in Peninsular Malaysia', *Population and Development Review* 18 (1992): 243–267.

[21] N. Wade, 'Sahelian Drought: No Victory for Western Aid', *Science* 185 (1974): 234–237.

[22] John F. May, 'Policied on Population, Land Use and Environment in Rwanda', *Population and Environment* 16 (1995): 322.

[23] V. Abernethy, 'Optimism and Overpopulation', *The Atlantic Monthly* (December 1994): 84–91.

[24] *Ibid*; Abernethy, 1993.

[25] J.C. Caldwell, I.O. Orubuloye and P. Caldwell, 'Fertility Decline in Africa: A New Type of Transition?' *Population and Development Review* 18 (June 1992): 229.

Africa provides a particularly clear example of the impacts of misdirected development assistance. The continent received three times as much foreign aid per capita as any other part of the world in the decades following World War II, and it now also has the highest fertility rates. During the 1950s and 1960s, the African fertility rate rose to almost seven children per woman, at the same time that infant mortality was dramatically reduced, health care availability grew, literacy for women and men became more widespread, and economic optimism pervaded more and more sectors of society. Algeria, for example, achieved independence in 1962, and 30 years later 70 per cent of its population was under 30 years of age. Zimbabwe gained independence in 1980 and soon achieved one of the highest population growth rates in the world.[9] Rwandan population growth was probably spurred both by independence in 1962, and by official policies that included dispersing people to virgin agricultural lands as well as into neighbouring countries, which likely conveyed to migrants a sense of fresh opportunity and liberation from spatial constraints. The fertility rate reached 8.6 births per woman by 1987.[10]

Other data also show that fertility rates follow perceived opportunity. If people of one ethnic group or class improve their access to opportunity at the expense of another within the same country, the shift is reflected in changing fertility rates. Similar patterns are observed with respect to immigration opportunities. For example in the US, while native-born populations are relatively stable, immigrant populations are experiencing employment and welfare opportunities far greater than those available in the countries they left behind, and this is reflected in fertility rates among immigrants that are higher than those of native-born members of their own ethnic group, and in some cases such as Mexico higher than those in the countries they leave behind. In addition, studies of nineteenth-century England and Wales and of modern Caribbean societies show that in communities already in the throes of rapid population growth, fertility stays relatively high as long as the option to emigrate exists, whereas fertility falls rapidly in communities that lack such an escape valve.[11] This is consistent with findings indicating that emigration raises incomes both among emigrants and among those they leave behind.[12]

In sum, it may be an uncomfortable truth that efforts to alleviate poverty often spur population growth. Subsidies, windfalls or prospects of unusual economic opportunity remove the immediacy of the need to conserve. The mantras of democracy, redistribution and economic development raise expectations and fertility rates, fostering population growth and thereby steepening a downward environmental and economic spiral. Development programmes that entail large transfers of technology and funds to the Third World are especially pernicious. This kind of aid is inappropriate because it sends the signal that wealth and opportunity can grow without effort and without limit; desired family size then rises, and rapid population growth ensues.

Some experts and members of the public nevertheless continue to believe that fertility rates have traditionally been high worldwide, and have declined only in industrialized countries or in countries where modern contraception is available. They argue that the post-World War II population explosion is explained mostly by better health and nutrition, which leads to rapidly declining mortality rates and slight, involuntary increases in fertility. The possibility that larger family sizes have resulted from a desire for more children continues to be denied.

These misperceptions date back at least to the 1930s, when many demographers predicted a steady decline in population because the low fertility of western industrialized countries was attributed to development and modernization, rather than to the endemic pessimism brought on by the Great Depression.[13] They failed to recognize that the high fertility beginning after World War II was a response to the perception of expanding economic opportunity. The US baby boom (1947–61), as well as the slightly later boom in western Europe, took most demographers by surprise and the causes continue to be misunderstood despite efforts to set the record straight.[14]

THE MESSAGE OF SCARCITY

As it happens, the encounters with scarcity that are currently being forced upon literally billions of people by the natural limits of their environment are beginning to correct the consequences of decades of misperception. Now, as at many other times in human history, the rediscovery of limits is awakening the motivation to limit family size (see Table 9.1).

For example, the population growth that occurred in post-revolutionary China continued until a famine in the late 1950s and early 1960s, which was unrelieved by Western aid, forced a confrontation with limits. In 1979, mindful of severe food shortages, the government instituted a one-child-per-family policy, thus completing the evolution of incentives and controls that returned the country to the pre-Communist pattern of marital and reproductive restraint.[15] Meanwhile, in the former Eastern-bloc countries, economic restructuring, the dissipation of government consumption subsidies, the public perception of rising infant mortality and a sense of scarcity or of 'relative deprivation' – that is, poverty in comparison to one's reference group – all promoted lower fertility rates.[16]

In the US, the baby boom ended at about the same time that economic opportunity levelled off and job opportunities became more scarce. The fertility rate dropped below replacement level after the 1973 oil shock occurred and many Americans' real incomes stopped rising. In the US today, one can surmise that the below-replacement-level fertility rate (1.8) among the native-born sector of the population results from the failure of actual economic opportunity to live up to earlier expectations.

Meanwhile, in the developing world, prodded by international economic retrenchment in the late 1980s, the government of Zimbabwe began to actively support family planning and it found the populace receptive. According to *The Economist*, 'The hefty cost of supporting a large family has helped persuade some men of the value of limiting its size'.[17] However, when there is no motivation to limit family size, providing modern contraception can be nearly irrelevant.[18] For six years in the 1950s, a project directed by British researcher John Wyon provided several villages in northern India with family planning education, access to contraception and medical care. The villagers had positive attitudes towards the health-care providers and towards family planning, and infant mortality fell considerably, but the fertility rate remained high. Wyon's group soon figured out why: the villagers liked large families. They were delighted that now, with lower infant mortality, they could have the six surviving children they had always wanted. The well-funded Wyon project may even have reinforced the preference for large families by helping to make extra children more affordable.[19]

THINK LOCALLY

Miscalculations about the causes of the population explosion have led to irrelevant and even counterproductive strategies for helping the Third World to balance its population size and its resources. In the late 1940s and the heady decades that followed, trade, independence movements, populist revolutions, foreign aid and new technology made people in all walks of life believe in abundance and an end to the natural limits imposed by the environments with which they were familiar. Now it is necessary for industrial nations, with their wealth much diminished, to reevaluate and target their aid more narrowly. The idea that economic development is the key to curbing world population growth rests on assumptions and assertions that have influenced international aid policy for some 50 years. These assumptions do not stand up to historical or anthropological scrutiny, however, and the policies they have spawned have contributed to runaway population growth, thus doing a disservice to every country targeted for development.

With a new, informed understanding of human responses, certain kinds of aid still remain appropriate, such as micro-loans that foster grassroots enterprise where success is substantially related to effort. Micro-loans to women seem particularly worthwhile because moving women into the cash economy by almost any means appears to lower their family-size target. This effect is independent of education – although it may reinforce the perceived need for education – because it operates through a simple calculation that anyone can make: the time spent in childrearing takes away from the time available for earning money. Childrearing entails opportunity costs.

Micro-loans are implemented, as a rule, through non-governmental agencies, but they can be appropriately funded through either government or private giving. This modest agenda remains within the means of industrialized countries even as they look to the needs of the growing ranks of their own poor, and it does not mislead and unintentionally harm intended beneficiaries.

Humanity's strength is its ability to respond quickly to environmental cues – a response most likely to be appropriate when the relevant cues come from the immediate, local environment, rather than a distant, external one. The mind's horizon is here and now. Our ancestors evolved and had to succeed in small groups that moved around relatively small territories. Not surprisingly, then, signals that come from the local environment are powerful motivators.

Let the globalists step aside: one-world solutions do not work. Local solutions will. People everywhere act in accord with their perceptions of their best interests. They are adept at interpreting *local* signs to determine their next move. As a result, in many countries and communities today where social, economic and environmental conditions are indubitably worsening, the demand for family planning is rising, marriage and sexual initiation are being delayed, and family-size is contracting. This is the local solution: individuals responding to signs of limits with low fertility.

Notes

1. This is an updated version of Virginia Deane Abernethy, 'Optimism and Overpopulation', *Atlantic Monthly* (December 1994): 84–91.
2. Shaukat Hassan, 'After UNCED: The Population Issue Revisited', as summarized in the conference report for the Third Conference on Environmental Security titled *Population/Environment Equations: Implications for Future Security*, Tufts University, Medford, Massachusetts, 31 May–4 June 1994.
3. V. Abernethy, *Population Pressure and Cultural Adjustment* (New York: Human Sciences Press, 1979); V. Abernethy, *Population Politics: The Choices that Shape Our Future* (New York: Plenum Press, 1993); V. Abernethy, 'Optimism and Overpopulation', note 1 above; V. Abernethy, 'Population Theory and Future Population Size', in *Environmental Pollution – ICEP3*, ed. B. Nath (London: European Centre for Pollution Research, 1996); W.R. Catton, Jr., 'The World's Most Polymorphic Species: Carrying Capacity Transgressed in Two Ways', *Bioscience* 37 (1987); W.R. Catton, Jr., 'Backing Into the Future', *Focus* 5 (1995): 41–6; and E. Goldsmith, 'The Population Explosion: An Inevitable Concomitant of Development', *The Ecologist* 19 (1989): 2–3.
4. See, for example, Paul Demeny, 'Social Science and Population Policy', *Population and Development Review* 14 (1988): 451–80. His discussion of this is exceptionally clear.
5. Lant A. Pritchett, 'Desired Fertility and the Impact of Population Policies', *Population and Development Review* 20 (1994): 34.
6. Population Reference Bureau, 'Man's Population Predicament', *Population Bulletin* 27 (1971): 19.
7. Judith Banister, *China's Changing Population* (Stanford, California: Stanford University Press, 1987), 3.
8. *Ibid.*, 233.
9. Abernethy, *Population Politics*, note 3 above.
10. John F. May, 'Policies on Population, Land Use, and Environment in Rwanda', *Population and Environment* 16 (1995): 322.
11. A.W. Brittain, 'Migration and the Demographic Transition: A West Indian Example', *Social and Economic Studies* 39 (1990): 39–64; A.W. Brittain, 'Anticipated Child Loss to Migration and Sustained High Fertility in an East Caribbean Population', *Social Biology* 38 (1991): 94–112; and D. Friedlander, 'Demographic Responses and Socioeconomic Structure: Population Processes in England and Wales in the Nineteenth Century', *Demography* 20 (1983): 249–72.
12. Abernethy, 'Optimism and Overpopulation', note 1 above.
13. See, for example, F. Notestein, 'Population: The Long View', in *Food for Thought*, ed. T.W. Schultz, Norman Wait Harris Memorial Lectures, 1945.
14. R. Easterlin, 'The American Baby Boom in Historical Perspective', Occasional Paper no. 79 (New York: National Bureau of Economic Research, 1962).
15. Banister, *China*, note 7 above.
16. C. Haub, 'World Population Expected to Reach 6 Billion in Early 1999', *Population Today* 24 (1996): 1–2; and C. Conrad, M. Lechner

and W. Werner, 'East German Fertility After Unification: Crisis or Adaptation', *Population and Development Review* 22 (1996): 331–58.

17. Abernethy, 'Optimism and Overpopulation', note 1 above.

18. B. Robey, S.O. Rutstein and L. Morris, 'The Fertility Decline in Developing Countries', *Scientific American* 269 (December 1993): 60–7; and Pritchett, 'Desired Fertility', note 5 above.

19. J. Wyon and J.E. Gordon, *The Khanna Study: Population Problems in the Rural Punjab* (Cambridge: Harvard University Press, 1971).

10 Population and Urbanization in the Twenty-First Century: India's Megacities[1]

Sai Felicia Krishna-Hensel

Megacity growth is rapidly becoming a central feature of India's transformation as it enters the next millennium. To many analysts of the region, this element of Indian urbanization reflects both economic progress, as well as processes of social change. Like other developing regions, south Asia views urbanization as a necessary part of development and a valuable index of growth.[2] The presence of these urban agglomerations is therefore seen by residents and administrators as evidence of modernization and economic opportunity. It is, however, becoming increasingly clear that the problems associated with growth will need to be addressed more effectively, both at the local and the national levels.

Megacity growth reflects the social transformations occurring in Indian society, including the emergence of a new kind of urban personality. Traditional family units and social groupings are realigning themselves in response to urban realities such as crime, pollution and poverty, problems that are magnified in megacities to proportions hitherto unimaginable. There is also growing evidence to suggest that alcoholism and prostitution are becoming more prevalent amongst the urban poor.[3] At the same time, however, there is a growing awareness that the change from a pre-industrial to an urban industrial society requires a complex response involving several facets of urban management. Municipalities and other government institutions must simultaneously cope with environmental, social, religious, cultural, economic and spatial issues.

The beginnings of India's transformation from a predominantly rural society to one in which 26 per cent of the population currently lives in urban areas can be traced to the events of the post-independence period.[4] Initially, the largest population concentrations were in those cities that had dominated the pre-independence urban scene, especially Calcutta, Bombay and Madras. The involvement of these cities in trade and commerce served as a magnet for individuals seeking employment opportunities. This pattern of continuing growth in already large cities is characteristic of south Asian urbanization. These dominant cities have now reached the category of megacities – urban areas with populations exceeding eight million – and in fact it appears that most of them will reach at least double that figure by the year 2000.

The phenomenon of megacity growth points to the historical importance of good location for the development of urban centres. These locations were selected in earlier times because of climate, transportation access, and/or availability of natural and manufactured resources,[5] and these features have not changed over time, continuing to attract migrants in search of opportunities. Economic patterns established in an earlier era do not appear to have changed either; the economic forces that drove the initial growth of cities at these sites continue to be as vital as before, though modified in response to changing needs so as to continue attracting investment and workers.

Megacities face a host of potentially overwhelming problems and management challenges. The magnitude of these urban problems has revealed the limitations of city governments' responses to the needs of their inhabitants, and points to the need to encourage the inclusion of non-traditional institutions in providing solutions. Megacities appear to be too large a unit to be efficiently administered through one central agency, warranting the delegation of considerable decision-making power – for example, over the provision of urban utilities and the maintenance of law and order – to local wards. Local wards in megacities might also have to be given power to collect revenues for special local needs, such as education or health and hygiene, and possibly even the authority to regulate the entry of new residents. Implicit in this approach is a recognition that local institutions should be created and put in place in anticipation of the future growth-related needs of urban populations.

The megacity is more than just an overwhelming challenge for administrators and residents, it is also the crucible of creative solutions and a laboratory for human ingenuity, as demonstrated by the numerous examples discussed in this article. Where official help is absent or slow in coming, residents have formed self-help groups to deal with their problems, and in many instances youth are being enlisted to spearhead urban sanitation, health and civic education programmes. The following examination of India's urban giants will evaluate the growth patterns of the principal cities in a historical context. Some of the social and environmental problems that characterize rapid unplanned urban growth will be used as indices of the changing nature of urban society. Finally, this study will assess the challenges faced by planning institutions and profile the emergence of non-traditional responses.

PATTERNS OF GROWTH

Factors favouring the initial siting of cities, such as locations that link inland transportation routes with external sea routes, and the presence of trading facilities that led to the development of coastal cities, continue to be relevant in explaining the growth of the modern Indian megacities. The expansion of processing and manufacturing capabilities has provided another incentive for this growth. The fundamental shift from a traditionally agrarian to an industrial economy – which has become central to the country's vision of its future as an independent nation – has served as an impetus for the population movement towards those cities most engaged in industrial production and trade. Economic forces favouring greater investment and growth in this part of the world will generate still more jobs, and thus large numbers of migrants will continue to be drawn to the cities, exacerbating existing shortages of land and other urban services.

The unprecedented swelling of urban populations in megacities does result in part from natural population increases, but large-scale migration is the predominant cause. A comparison of the migration patterns of four of the principal cities – Calcutta, Bombay, Madras and Delhi – does, however, show significant variation among them, reflecting the diversity of factors that draw migrants to the cities. Economic opportunity and industrial productivity explain part of the attraction, as migrants are attracted by the job opportunities in the extensive small-scale manufacturing units. In addition, rural populations flock to the cities seeking skills and education.

The ranks of urban migrants have been further increased by two important historical events: the partition of the sub-continent in 1947 that led to the creation of India and Pakistan, and the partition

of Pakistan in 1971 leading to the creation of Bangladesh out of
the former East Pakistan.[6] In both instances, extensive displacement of the population led to large-scale refugee movements that
enlarged the populations of cities such as Calcutta. As with most
sudden influxes of large masses of people, the civic authorities in
Calcutta were generally unprepared to respond to this crisis, and
housing, medical attention, food, water and sanitation were all in
short supply. As a result, refugees resorted to pavement dwelling –
Calcutta is probably the city most identified with pavement residents – creating a major health and safety hazard in the city. The
creation of the Calcutta Metropolitan Planning Organization (CMPO)
was an effort to provide standardized guidance to the municipal
authorities on spatial management and public-works projects.[7]

The population profile of Bombay is also strongly influenced by
migration. However, in contrast to Calcutta, where many migrants
came as refugees, in Bombay migrants generally converged on the
city from surrounding areas both within the state and in three neighbouring states (Mysore, Gujarat and Uttar Pradesh) in response to
new employment opportunities. Although one would assume that
in the absence of sudden influxes of migrants civic authorities would
be better prepared to handle the situation, Bombay has the distinction of having nearly half of its population residing in squatter
or slum housing, with all of the attendant problems that this entails.[8]

The growth of Madras reflects in-migration and natural increase
in almost equal measure. Most migrants are rural residents of the
state drawn to the city by economic opportunity. The number of
migrants from other states is not substantial, and these migrants
also tend to be in the middle to upper income category, whilst in-
state migrants are generally in the lower income groups.[9] Even so,
the proportion of slum housing in this city is relatively small compared with the other megacities of the sub-continent.

The nation's capital, Delhi, experienced a massive influx of refugees following partition in 1947, but a substantial proportion of
this population was more fully absorbed into the city during the
following years. Government-subsidized housing colonies were constructed to accommodate the residents of temporary tent cities, so
there are few truly abject examples of slum housing within the city
boundaries. Delhi was also one of the first megacities to recognize
that the success of any future attempts at planning would have to
be based on regional planning efforts. As a result, the National
Capital Region Board (NCRB) has consistently emphasized dispersal

of growth into the surrounding townships of Faridabad, Ghaziabad, Gurgaon and others. Relying on indicators such as 'carrying capacity of the regional environment', Delhi planners have sought to avoid the mistakes of the other megacities of India.[10]

URBAN PROBLEMS

Megacities face many problems and challenges ranging from pollution to social problems. Housing is one of the most serious areas of concern as the rise of shanty towns constructed of flimsy materials has become a central feature of urbanization. These hastily constructed homes obviously do not follow any building codes, representing fire and health hazards that have repercussions beyond their immediate location. Overcrowding likewise takes a toll on privacy, as well as safety.

Providing services in these areas is especially difficult. Electricity is often pirated from existing power lines by those who can afford to do so. The makeshift shelters are unauthorized by municipal authorities, making it difficult to locate water and sewer lines and make connections. Residents of squatter settlements often illegally tap into public hydrants to obtain a minimal water supply. Untreated sewage also proves to be hazardous, especially during the monsoon season when heavy downpours wash it into the general water supply, a situation reflected in the growing incidence of disease. Nor do squatter settlements have access to authorized schools or health-care facilities. Some unauthorized enterprises do provide minimal education and health care, but in the absence of regulation they often exploit the public. An exception is the UNICEF project for improving the nutritional quality of infant feeding in Delhi, which hopes to establish the importance of early intervention in preventing malnutrition. This has been described as the 'world's largest public feeding scheme', with an annual budget of $350 million.[11]

The crowded conditions that prevail in the squatter settlements are a breeding ground for pollution and environmental degradation, disease, delinquency and crime, and the impacts are felt throughout the city, decreasing the quality of urban life for everyone. Based on government data, it appears that between 40 and 70 per cent of residents in cities like Calcutta, Bombay and Madras reside in slums or overcrowded tenement housing.[12]

High land prices, air and water pollution, and a growing climate of corruption and extortion are other features of megacities that have negative impacts on everyone. The cost of land, especially land in the city centre that is regarded as prime real estate, has been appreciating at a dramatic pace. This increases the potential for corruption and illegal profiteering, endangering the orderly development of these cities. Legal impediments such as the Urban Land Ceiling and Regulation Act of 1976 are a major deterrent to progress, as the provisions protecting workers' jobs and the high capital gains tax rate prevent landowners from developing their own properties. As a result, estimates of office space rental prices range from $600 per square metre in Delhi to a high of $900 per square metre in Bombay.[13] Bombay best exemplifies the enormity of the challenge facing the authorities. The entry of gang-related elements into the real estate market has led to extortion, bribery and even murder of landowners, as exemplified by the brutal murder by gang members of a Bombay textile magnate whose factories occupied prime real estate, but were in need of modernization. Underworld crime figures realized the potential profitability of a situation in which the factory could be relocated in response to state government incentives encouraging mill owners to relocate to cheaper land on the outskirts of the city, and the land could then be sold for redevelopment for a handsome profit.[14]

The practice of extortion has become widespread in Bombay. All new building or remodelling is reportedly subject to a 10 per cent levy by crime syndicates, and bribery of politicians and members of the city administration ensures that the authorities will have no incentive to intervene effectively. It is also estimated that large numbers of poorly paid police are entering the world of syndicated crime.[15] In New Delhi also, real estate values have been rising in response to demand for land in the central areas of the city, fueling widespread corruption among city officials. Diplomats in need of housing and foreign investors looking for prime real estate to locate their companies are particularly vulnerable to this development. Renters are generally those who control enough income to afford such payments, but who are not wealthy enough to own their own residences and real estate in this environment of rapidly escalating urban land prices. Ironically, the poor who live in their own structures in urban slums do not have to concern themselves with the payment of rent. Meanwhile, many renters are forced to live in derelict and unsafe housing as landlords neglect their properties in the hope that when the structures collapse they can sell the land to developers for exorbitant prices.[16]

Urban air pollution has become a major health hazard in cities like Delhi. The effects are not only physically visible – most cities are covered by a haze of pollution – but they can also be seen in the rising rate of respiratory disease among the urban population. Once again, Delhi leads in this respect, with estimates that 30 per cent of the city's inhabitants are severely affected by air pollution, a figure that is far in excess of the national average.

The sources of air pollution range from vehicular emissions to combustion of fossil fuels.[17] Economic prosperity has led to a dramatic increase in the number of two-wheeled vehicles in the city, resulting in heavy traffic and increasing emissions of untreated hydrocarbons.[18] In Delhi alone, over 700 new vehicles are reportedly introduced into traffic every day.[19] World Health Organization limits are exceeded for several pollutants, including suspended particulates, and at present rates of increase other air pollutants such as nitrogen oxides will also exceed global guidelines in the near future. Another major factor contributing to deteriorating air quality is the increasing production of smoke by a growing number of households using biomass fuels and wood for cooking.[20] Moreover, the source of wood is often surrounding forest areas, which are rapidly shrinking due to increasing demand for wood and expansion of the urban area. This in turn reduces the green cover that can absorb some of the components of air pollution produced in the cities.[21]

The reliability of air quality data has been questioned by analysts since most of it is generated by a single agency, the National Environmental Engineering Research Institute (NEERI) through its national network, and no corroborative information is available from other sources. NEERI maintains one or more monitoring station in most of the large cities. Each collects data on at least four pollution categories – suspended particulates, nitrogen dioxide sulfur dioxide, and ammonia – at regular intervals (on an hourly or daily basis, depending on the city and need). At present, however, there is no systematic collection of information on ozone levels or carbon monoxide emissions. A new programme, the National Ambient Air Quality Monitoring Programme, was initiated in 1985 under the auspices of the Central Pollution Control Board, but differences in criteria, methodology and siting have resulted in anomalies that make it difficult to compare data from these sources with any measure of reliability.[22]

Water supply is already a serious problem in India's cities, with demand far exceeding supply. Most cities resort to restricting the hours of supply, especially during the summer months when rainfall is scarce and surface evaporation is high. In addition, access to filtered and treated water is uneven, with squatter colonies suffering the most. Despite modernization of treatment plants, replacement of outmoded pipes and digging of wells, both the quality and the quantity of water available leaves much to be desired.

In the area of sewage treatment there is a similar imbalance between public needs and the cities' ability to meet them. A comparison of the percentage of sewage that is generated by the urban population and the percentage that is treated reveals some alarming inadequacies. In the nation's capital, approximately 20 per cent of the sewage receives full treatment and about 30 per cent is partially treated, leaving roughly half of the city's sewage totally untreated.[23] This is typical of conditions elsewhere as well. Solid waste, meanwhile, is disposed of manually, along with animal waste products from both stray urban cattle and small dairies located inside the urban area. Most of these wastes are dumped outside the city in untreated pits, thereby posing a health hazard which is further compounded by tropical conditions, especially heat and flooding rains.

Attempts to correct these conditions have included increased investment in construction of additional treatment plants, and replacement of obsolete pipelines and pumping systems. Assistance from the World Bank has enabled cities like Bombay and Calcutta to almost double their investment in sewage management projects, although Calcutta still does not have a mainstream sewage treatment facility.[24] Even Delhi falls short in this area, with over 1500 tons of uncollected garbage cluttering the city streets. As Table 10.1 illustrates, the major cities generate more garbage than the municipalities are capable of removing, and they discharge sewage in amounts that far exceed the available treatment capacity. If these figures are an accurate assessment of reality, then anywhere from 10 to 37 per cent of garbage remains uncollected every day, while 4 to 30 per cent of sewage is not treated. Delhi appears to lead in environmental neglect.

Local rivers already reflect high levels of pollution from releases of untreated sewage, and the rapid influx of migrants into the cities means that large numbers of new households in outlying areas will not be provided with municipal pipelines either. The sanitation

Table 10.1   Daily garbage and sewage generation and disposal

	Garbage output (metric tons)	garbage collected (metric tons)	Sewage output (million litres)	Sewage treated (million litres)	City budget (crores)
Delhi	3880	2420	1800	1260	1016
Bombay	5800	5000	1800	1460	2436
Madras	2675	2140	250	238	250
Calcutta	3500	3150	800	675	145

Source: *India Today* (11 October 1994): pp. 40–1.

problems of the megacity will continue to be a major challenge as rapid expansion strains the cities' capacity to cope with the influx of migrants. The answer may ultimately lie in finding creative community-based solutions and decentralizing some of the implementation of these municipal functions to local groups. A model of community-based initiatives is the Exnora International organization, which reportedly collects about 20 per cent of the garbage generated in Madras for a modest fee. The fee varies based on the numbers of residents served, with larger neighbourhoods receiving lower per capita assessments. Privatization of this nature may be the ultimate answer for overburdened municipalities.[25]

The grim picture presented by current water and sewage disposal problems is relatively mild in comparison to what may lie ahead for these cities if widespread disease and pestilence break out. A foretaste of such a situation was experienced by Indian authorities in 1993 when a reported outbreak of plague in northern India served as a wake-up call. Even though cases of cholera, diarrhoea, gastroenteritis and malaria number in the millions each year, the panic and fear unleashed internationally by the plague outbreak was unprecedented in recent times. Municipal authorities in all of the major cities were forced to take note of the appalling conditions that prevail in most urban areas, and they began to implement new sanitation measures and more regular pickup of garbage, while launching popular campaigns to increase public awareness of sanitation issues.[26]

Recognizing that one of the greatest threats to urban civilization is the rapid spread of pestilence and disease in an overcrowded environment is a first step in a long process. Planners are also recognizing that despite the advances of modern medicine, disease can only be controlled if all residents cooperate in immunization

programmes, observe sanitary laws, and take responsibility for reducing pollution at the individual and local community levels. This
clearly represents a new attitude towards problem-solving and a
newly-discovered sense of strength at the grass roots level. 'The
awakening of self-reliance in the urban poor is a global phenomenon'.[27] International agencies involved in providing aid have also
been forthcoming with higher levels of support for community
projects. An unanticipated outcome of this growth has been the
emergence of creative grassroots movements among the residents
of endangered communities themselves.

THE PLANNING CHALLENGE

Solutions are often only as effective as the careful planning that
precedes them, and this in turn is largely dependent on the availability of accurate data. India's urban planners and administrators
have relied on a variety of sources to provide adequate information for policy formulation. Traditional sources, especially the national
census, have been used extensively since independence. One problem has been that the national census, taken every ten years, only
reflects changes long after they have started to have an impact on
the urban situation. As a result, burgeoning birth rates and declining death rates, in combination with a number of unanticipated
factors ranging from natural disasters to political upheavals, have
prevented policy-makers from effectively coping with the rapid growth
of cities. Urban management based on standard census data cannot
be sensitive to the complexities of megacity challenges.

Calcutta provides a prime example of this problem. The population of the city doubled from 4.4 million in 1950 to 9 million in
1980, a period of just 30 years, and during the next 10 years accelerated growth rates added another 2.8 million inhabitants. The city
is projected to have 15.7 million inhabitants by the year 2000.[28]
Contributing to this growth profile has been the creation of a new
nation bordering West Bengal in 1973 which led to an influx of
political refugees at that time. Meanwhile, improved medical attention and the effects of the 'green revolution' were altering prior
patterns of high fatalities due to malnutrition and disease. Relying
on standard population data, policy-makers were inadequately prepared for these rapid changes. In addition, much of the data is subject to dispute as there is substantial variation between government

statistics and those from other sources. Poverty figures, for example, range from a low of 18 per cent to a high of 37 per cent, thus calling into question any speculation on the effects of economic growth.[29]

The imprecision that characterizes the definition of terms such as urban, rural, suburban and satellite townships further complicates the task of policy-makers. These terms have different connotations in different societies, requiring a close look at the manner in which standard terminology is formulated. For example, the growth of commuter traffic into the cities has begun to obscure the distinction between permanent and temporary residents. It is not uncommon for individuals to maintain a technical residence inside the city, usually at a relative's or friend's address, while actually residing in a nearby township. This greatly complicates the problem of obtaining an accurate population count for tax collection and provision of services. City boundaries are also being redefined frequently to accommodate outward expansion, often absorbing outlying villages.

Notwithstanding the need for a more comprehensive data collection system, as well as more frequent updating, the currently available statistics point to a clear trend of increasing megacity growth in the region. The impact of this unwieldy urban phenomenon on the surrounding areas is as yet unknown, but it is clear that the new urban culture that is emerging in these cities is complex, with new economies forming that are more decentralized and organized along informal lines. Policy-makers are confronting urban problems of an unprecedented – and still growing – magnitude, but they are finding assistance in unusual places, including decentralization, privatization of services and community-based support groups, all of which can help to sustain the quality of urban life and provide basic services.

By their very nature, megacities appear to be beyond the reach of standard social or physical planning methods. Within this seemingly impossible situation, however, lie some of the possible solutions to the management problems of these cities. In the first place, more of this urban expansion is now being planned as developers scramble to provide housing for the ever-growing population. However, this generally benefits those who can afford the prices, and excludes the much larger numbers of urban poor. Municipalities thus need to take a more aggressive role in demanding that new developments provide a range of services to the residents at reasonable prices. In addition, granting development permits for upper-income residents should be linked to additional developer contributions levied by the municipalities to provide some of the missing services in low-income, unplanned neighbourhoods. This cooperation between municipalities and developers would go a long way towards reducing the glaring gap between urban slums and suburban areas. Concerted efforts by the public sector and private enterprise could improve conditions for all urban residents.

Planners have developed a number of strategies for specific cities. Bombay, for example, has adopted the concept of a twin city to facilitate redistribution of the population. This concept involves developing a number of small townships, some of which are already in place and others that are scheduled for construction. This proposal includes provision of low-income housing to a large segment of the urban population, and early plans also call for developing space for offices and commodities markets in an attempt to encourage relocation of businesses and thus relieve congestion in the old city.[30] Calcutta planning authorities are also trying to decentralize urban activities by developing several surrounding townships with the aim of relieving congestion in the central city. Planners are also trying to rechannel the city's natural growth trend away from the northwest and towards the southeast. These efforts are eventually expected to lead to the decentralization of several of the city's functions, including relocation of government agencies and development of port facilities.[31] Madras has followed suit by encouraging the relocation of city agencies and commodities markets to outlying developments. In addition, the city is trying to institute re-use patterns that will ensure that industries and activities requiring a smaller workforce will relocate in the vacated areas.[32] Delhi's future is being planned to include a National Capital Region that encompasses several outlying districts and counter-magnet cities – that is smaller urban townships that are expected to grow.[33]

The distribution of a city's population is determined in part by both the location of jobs and the time of day. For example, if the central city is the locus of administrative and business offices, there will be a large influx of temporary workers during the workday, who will return to their residences at its close. During the workday, vehicular and human congestion increases, pollution levels are elevated, and municipal services such as water, electricity and transportation will face heavy demand. Data on the spatial distribution of jobs indicates that approximately one million workers arrive in the central city in Delhi each day, while only about 40 000 enter Madras for this reason. It is unclear from the available data whether demands for services increase in residential sections of the city at the end of the workday as the population is redistributed for the night hours. If businesses and administrative offices are encouraged to relocate to outlying areas presumably new patterns of usage will emerge, but at present there are no reliable studies to help planners and city administrators effectively respond to the changing needs of shifting populations.

The pattern of rapid urban population growth has revealed the limitations of the traditional mechanisms of urban management which, with their heavy reliance on census data have been unable to react to rapidly changing circumstances. Census data does not reflect the continual shifts that characterize the growth of megacities, in part because data-collection is too infrequent to reflect these trends. The absence of reliable migration statistics is not unique to India; as noted in the report of the 1984 United Nations International Conference on Population, more emphasis needs to be placed on differentiating among data on external migration internal migration, and urban migration.[34] The current database has many gaps, for example with respect to the weights assigned to reasons for migration (ranging from economic or educational opportunities to marriage or the desire to join family members). Each of these reasons has different consequences for the tax base, the use of public services, the degree of identification with the community, and long-term policy formulation. Presently, however, there is no effective way to keep databases updated to reflect changing patterns.

An area that has not been fully explored in the sub-continent is the harnessing of technology to improve the quality of life in megacities. For example, Calcutta has the only subway presently providing transportation between peak traffic areas, but in future surface congestion will have to be dealt with by the introduction of a more extensive underground transportation network. This would yield benefits both by reducing pollution, and by reducing the amount of time spent in transit between the home and the workplace, which would generate more hours of productivity and yield more family time for urban workers.

The urban food supply is another problem area that has yet to receive systematic attention. The uneven distribution of food in cities is not only a function of the purchasing power of the local population, but also of the inefficiency of the supply system. As cities expand into the rural hinterland, farmers are required to transport food across longer distances while central produce markets become congested and inaccessible to suburban consumers. Cities are trying to relocate some of their markets to outlying developments, and informal kitchen gardening is continuing to flourish in many areas where vacant or unclaimed property is being temporarily converted by private entrepreneurs into dairy farms and vegetable gardens. However, the uncertainty of such enterprises, which can only function as long as no-one challenges the operator, does not commend them as a successful or even desirable solution to the problem of feeding the megacity population.

In conclusion, the problems of megacities present monumental challenges to policy-makers in India, but there are promising avenues that may help reduce the gravity and scope of these challenges. The future of megacities will be determined in large part by the ability of planners to accurately assess growth trends and to prepare workable solutions for the problems that will inevitably ensue. Accurate assessment is dependent on reliable data-collection, which is particularly important for pollution measures. Cooperative ventures between non-governmental bodies and municipal agencies will be required to address the magnitude of the urban challenges confronting the megacities. Decentralization and extra-urban development is likely to be the best direction for the future. In addition, improved transportation technology, currently an underused resource, may ultimately prove to be the answer to the problem of urban congestion. Finally, the importance of education in mobilizing popular support for dealing with the challenges of the future cannot be over emphasized.

Notes

1. The material in this chapter was first presented at the Thirty-Seventh Annual Convention of the International Studies Association, San Diego, California, 16–30 April 1996.
2. Dinesh Mehta, Usha Raghupati and Rajesh Sharma, 'Urban Ecology: Squalor That Need Not Be', *The Hindu Survey of the Environment* (1994): II-7.
3. 'The Fashion for Going Dry', *The Economist* (22 June 1996): 38.
4. *World Resources: A Guide to the Global Environment, 1994–95.* A Report by the World Resources Institute in collaboration with the United Nations Environment Programme and the United Nations Development

Programme (New York: Oxford University Press, 1994), 87 (hereafter cited as *World Resources: 1994–95*).

5. Sharada Dwivedi and Rahul Mehrotra, *Bombay: The Cities Within* (Bombay: India Book House, 1995), 16.
6. Percival Spear, *A History of India,* Vol. II (Harmondsworth: Penguin, 1970), 236–7.
7. Tridib Bannerjee and Sanjoy Chakravorty, 'Transfer of Planning Technology and Local Political Economy: A Retrospective Analysis of Calcutta's Planning', *Journal of the American Planning Association* 60 (Winter 1994): 4–5.
8. United Nations Department of International Economic and Social Affairs, *Population Growth and Policies in Mega-Cities: Bombay*, Population Policy Paper no. 6 (New York: United Nations, 1986), 1 (hereafter cited as United Nations, *Bombay*).
9. United Nations Department of International Economic and Social Affairs, *Population Growth and Policies in Mega-Cities: Madras*, Population Policy Paper no. 12 (New York: United Nations, 1987), 2 (hereafter cited as United Nations, *Madras*).
10. Tridib Bannerjee, 'The Role of Indicators in Monitoring Growing Urban Regions: The Case of Planning in India's National Capital Region', *Journal of the American Planning Association* 62 (Spring 1996): 222–3.
11. 'India's Children: Ill Fed in Ignorance', *The Economist* (8 June 1996): 39.
12. United Nations Department of International Economic and Social Affairs, *Population Growth and Policies in Mega-Cities: Calcutta*, Population Policy Paper no. 1 (New York: United Nations, 1986), 1 (hereafter cited as United Nations, *Calcutta*); United Nations, *Bombay*, 1; United Nations Department of International Economic and Social Affairs, *Population Growth and Policies in Mega-Cities: Delhi*, Population Policy Paper no. 7 (New York: United Nations, 1986), 25–26 (hereafter cited as United Nations, *Delhi*); United Nations, *Madras*, 1.
13. Hamish Macdonald, 'Squatters Rights', *Far Eastern Economic Review* (30 June 1994): 40–1.
14. M. Rahman, 'Land Wars', *India Today* (30 June 1994): 40–4.
15. *Ibid.*
16. D.B. Gupta, *Urban Housing in India*, Staff Working Paper no. 730 (Washington, DC: World Bank, 1985).
17. Louise Williams, 'Asia's Urban Meltdown', *Sydney Morning Herald*, cited in *World Press Review* 41 (February 1994): 46.
18. *World Resources: 1994–95*, 94.
19. 'The Degradation of Delhi', *The Economist* (4 December 1993): 36.
20. Kirk R. Smith, *Biofuels, Air Pollution and Health: A Global Review* (New York: Plenum Press, 1987), 25, 41– 57 and 207–9.
21. 'Air Pollution in the World's Mega-Cities: A Report from the United Nations Environment Programme and the World Health Organization', *Environment* 36 (4 March 1994): 4.
22. *Ibid.*
23. United Nations, *Delhi*, 27.
24. United Nations, *Calcutta*, 25.

25. Nirupama Subramaniam, 'Showing the Way', *India Today* (31 October 1994): 47.
26. 'Our Filthy Cities: Can We Clean Up the Mess?' *India Today* (11 October 1994): 37.
27. Eugene Linden, 'Mega-Cities', *Time* (11 January 1993): 28–38.
28. *World Resources: 1994–95*, 87.
29. 'How Poor is India?' *The Economist* (13 April 1996): 30.
30. United Nations, *Bombay*, 20–1.
31. United Nations, *Calcutta*, 18–20.
32. United Nations, *Madras,* 15–7.
33. United Nations, *Delhi*, 21–2.
34. Sidney Goldstein, 'Demographic Issues and Data Needs for Mega-City Research', in *Mega-City Growth and the Future*, ed. Roland J. Fuchs, J. Ellen Brennan, Joseph Chamie, Fu-Chen Lo and Juha I. Uitto (Tokyo: United Nations University Press, 1994), 32–61.

References

Benjamin, S.J. (1991) *Jobs, Land, and Urban Development: The Economic Success of Small Scale Manufacturers in East Delhi, India* (Cambridge, Mass: Lincoln Institute of Land Policy).
Brennan, E.M., and H.W. Richardson (1989) 'Asian Mega-City Characteristics, Problems, and Policies', *International Regional Science Review* 12: 117–29.
Cheema, G.S. (ed.) (1993) *Urban Management: Policies and Innovations in Developing Countries* (New York: Praeger).
Gappert, S. (1993) 'The Future of Urban Environments: Implications for the Business Community', *Business Horizons* 36 (November/December): 70.
Hamer, A.M. and J.F. Linn (1987) 'Urbanization in the Developing World: Patterns, Issues and Policies', in E.S. Mills (ed.), *Handbook of Regional and Urban Economics*, vol. 2 (New York: North-Holland).
Kingsley, G.T., J.P. Telgarsky and B. Walter (1989) *India's Urban Challenge: Trends and Implications* (Washington, DC: The Urban Institute).
Linn, J.F. (1983) *Cities in the Developing World: Policies for their Equitable and Efficient Growth* (New York: Oxford University Press).
Mills, E.S. and C. Becker (1986) *Studies in Indian Urban Development* (New York: Oxford University Press).
'New Homes for Old', *The Economist* (29 June): 33.
Oberai, A.S. (1989) *Problems of Urbanization and Growth of Large Cities in Developing Countries: A Conceptual Framework for Policy Analysis*. Population and Labour Policies Programme Working Paper no. 169 (Geneva: International Labour Office).
Rondinelli, D.A. (1983) *Secondary Cities in Developing Countries: Policies for Diffusing Urbanization* (Beverly Hills: Sage).
'The Most Expensive Slum in the World', *The Economist* (6 May): 35–6.
United Nations (1989) *Prospects of World Urbanization* (New York: United Nations).

United Nations Center for Human Settlements (1987) *Global Report on Human Settlements* (Oxford: Oxford University Press).

United Nations (1979) *World Demographic Estimates and Projections, 1950–2025* (New York: United Nations).

United Nations (1989) *World Population Trends and Policies – 1989 Monitoring Report*. Population Studies no. 103 (New York: United Nations).

World Bank (1990) *World Development Report 1990: Poverty* (Washington, DC: World Bank).

11 Mexico City: Current Demographic and Environmental Trends

Haydea Izazola and
Catherine M. Marquette

While increased urbanization remains the dominant demographic trend in Latin America, urban growth slowed in several of the region's major cities during the 1980s, including Mexico City.[1] Mexico City's highest growth rates (on average 5 per cent per annum) occurred from the 1940s to the 1960s. During this period, the city was transformed from an urban area of just over a million inhabitants to a megacity of 15 million. The city has grown to encompass both the core Federal District, which is administratively treated as a separate state, and 27 urban municipalities in the surrounding central state of Mexico.

Census data from 1990 indicate a possible turnaround in the city's long-term growth trend and in internal migration patterns, suggesting that Mexico City's power of attraction is declining. Although the urban area has continued to expand, the population growth rate of the city declined throughout the 1980s to less than 1 per cent per annum,[2] which is less than the rate of natural increase (the difference between births and deaths) of 2 per cent per annum. This reflects the fact that the country's main internal migration flows, which prior to the 1980s were primarily directed *towards* Mexico City, had changed dramatically by 1985, and internal migration was increasingly characterized by diverse patterns of *out*-migration from the city.[3] The city's net migration rates in the late 1980s were negative for the first time in this century.[4] Between 1985 and 1990, over 500 000 persons out-migrated from Mexico City, while less than 500 000 in-migrated, a net loss of over 100 000 people,[5] or approximately 0.6 per cent of the city's 1990 population of 15 million.

A confluence of factors have led to the decreased growth and the proliferation of out-migration flows from Mexico City between 1985 and 1990, and economic and industrial changes have been important factors driving these flows. Adverse changes in the environment of Mexico City, or ecological factors, may also have played some role. The discussion below further explores the interrelated industrial, economic, ecological and demographic trends affecting Mexico City in recent years.

INDUSTRIAL AND ENVIRONMENTAL TRENDS IN MEXICO CITY TO 1990

The current physical environment of Mexico City is the product of its past industrial, economic and natural history. From the 1940s to the 1980s, Mexico was characterized by an overwhelming concentration of industry in Mexico City. This concentration was consistent with the dominant pattern of development in Latin America after World War II, which focused on industrialization through import substitution. Consumer-goods industries were concentrated in primate cities, such as Mexico City, which offered an established market for consumer goods.[6] By 1970, Mexico City alone produced approximately half (47.0 per cent) of the country's gross industrial domestic product.[7]

This historical pattern of centralized industrial development abruptly ended during the 1980s.[8] Although absolute industrial production in Mexico City continued to rise, its relative industrial activity (indicated by its contribution to the country's gross urban manufacturing product) fell sharply during the late 1980s. In addition, the contribution of the city's core area (the Federal District) to the population economically active in industry also fell by one-third between 1980 and 1990.[9]

The contraction of industrial activity in Mexico City during the 1980s was the result of several major economic changes. Interest payments linked to the country's external debt soaked up financial resources, diverting them away from further investments in industry.[10] The country's adherence to the General Agreement on Tariffs and Trade (GATT) also resulted in a reorientation of government policy from import substitution towards an open economy. Consumer-good-imports flooded the domestic market, and domestic industries, which were concentrated in Mexico City, were particularly depressed. Increased land scarcity and rising operating costs also made Mexico City less attractive to industries during the 1980s.

The historical concentration of industry in Mexico City during previous decades has, however, played a major role in shaping the city's environment, particularly in relation to air quality.[11] From 1950 to 1980, there was little control over industrial air emissions. Industrial activity, combined with growth in the use of motor vehicles, led to the increased production of air pollutants, including sulphur dioxide, ozone,[12] carbon monoxide and suspended particles. Within the city, industrial development has been further concentrated in the northern and eastern parts of the metropolitan area from which the prevailing winds blow. As a result emissions generally drift south over the main metropolitan area. Wind erosion due to rapid land clearing for industry and housing has also produced high levels of suspended particles, another major source of air pollution. The economic development of the city that has accompanied industrialization has also stimulated increased levels of energy consumption and motor vehicle use which have further contributed to air pollution.

The above causes of air pollution have synergistically interacted with the city's geographic setting to create particularly severe pollution problems. Mexico City is naturally subject to higher dust and suspended particle concentrations due to the natural erosion of the ancient system of lake beds on which the city rests.[13] Its location within a valley also results in weak wind patterns, resulting in insufficient dispersion of air contaminants. This valley location also predisposes the city to thermal inversions in which the lower layer of warmer air, where contaminants are most heavily concentrated, fails to rise or disperse in the upper atmosphere and remains trapped over the city. Additionally, the city's high altitude (2240 metres above sea level) is associated with greater amounts of ultraviolet radiation, which enhances the conversion of primary air pollutants into secondary pollutants such as ozone. Also, at this altitude motor vehicles function less efficiently and produce greater amounts of polluting carbon monoxide and hydrocarbons.[14]

In the mid-1980s the government attempted the first comprehensive measures to address environmental conditions in Mexico City, focusing on air pollution. These measures have included the substitution of oil for natural gas in thermoelectric plants (which produces less contaminants), the closure and relocation of highly contaminating industries, car inspections, the reduced use of leaded gasoline, and the institution of a programme of non-circulation days for motor vehicles. In 1984, the government also began to comprehensively monitor air pollution in the city through an automatic monitoring network (RAMA) of 25 stations.

These measures have been judged to be generally ineffective.[15] Ironically, more stringent emission regulations in Mexico City may have encouraged polluting industries to relocate to other states and to replicate their negative impacts in other cities.[16] Moreover, it appears that since the late 1980s, air quality has continued to deteriorate in Mexico City despite these government measures. Emissions of carbon monoxide, ozone precursors (nitrogen oxide and hydrocarbons) and suspended particles have increasingly exceeded the limits defined as safe for human health.[17] Industrial contraction during the 1980s appears not to have led to any decline in polluting emissions or improvements in air quality. Any environmental benefits of industrial contraction may have been offset by decreased investments in the city's infrastructure and services.[18]

DEMOGRAPHIC AND ENVIRONMENTAL TRENDS IN MEXICO CITY TO 1990

Industrial concentration in Mexico City during the postwar period coincided with rapid rural-to-urban migration and demographic concentration. As noted above, the city's demographic growth rate for most of the period before 1980 was over 5 per cent per annum. Concurrent with this demographic growth, the country underwent a process of metropolitanization and was transformed from a largely rural to a primarily urban nation. More than half (56.2 per cent) of the entire population resided in urban areas by the 1980s.[19]

This previous era of rapid in-migration has been the major factor shaping population–environment relationships within Mexico City, and sociospatial segregation of the city's population is one major characteristic of these relationships.[20] From the 1940s to the 1970s, squatter settlements mushroomed on the city's periphery causing the physical limits of the urban area to expand rapidly. As a result, the largest proportion of the city's population, comprised of lower-income groups, is concentrated in the northern and eastern portion of this expanding periphery along with many of the city's industrial enterprises. A major part of the city's population is, therefore, directly and continually exposed to the highest concentrations of industrial emissions. The rapid expansion of the urban area via peripheral squatter settlements has also led to rapid changes in land-use, accompanied by high rates of soil erosion and suspended particle production, thus further affecting air quality in these areas.

Population growth and distribution in Mexico City has also been characterized by larger distances between places of residence and work sites. This has lead to the greater use of motor vehicles and higher levels of traffic, noise and air pollution. The city was also hit by a major earthquake in 1985 which destroyed a large portion of the city centre and caused the relocation of residences and government activity to the southern periphery. In addition to the immediate physical destruction caused by the earthquake, the new demands placed on infrastructure in the south also precipitated increased pollution in that area.

The rapid rate of urban population growth in all areas of Mexico City over the last three decades, as well as the city's overall population size, have made it impossible for urban infrastructure to keep pace with needs. Water and sanitation services and solid waste disposal are neither sufficiently extensive nor efficient, particularly in peripheral areas. In the face of this situation, the impact of the urban environment on population welfare, as well as the impact of population on the urban environment, must be considered. Although a lack of epidemiological data has been noted,[21] adverse health impacts (for example, respiratory, skin and eye problems and water-borne infectious disease) associated with the city's physical environment have been identified throughout the 1980s.[22] The recent contraction of industry and the concurrent decline in investments in already insufficient city services has probably only enhanced these negative health impacts.

THE ECOLOGICAL DETERMINANTS OF RECENT OUT-MIGRATION

As noted above, a proliferation of out-migration flows from Mexico City has occurred alongside industrial contraction and continuing deterioration in environmental conditions in the city during the 1980s. These out-migration flows may be largely directed towards median-size cities with populations of 500 000 to 1 000 000 in other parts of central Mexico. Results from the country's 1987 National Urban Migration Survey (ENMAU) indicate that the proportion of out-migrants from Mexico City moving to other cities in central

Mexico increased during the 1980s.[23] In contrast to Mexico City's lower growth, these median-size cities in central Mexico have grown at 6 to 10 per cent per annum during the last decade.[24] According to the 1987 ENMAU, those cities in central Mexico which have received the largest number of out-migrants from Mexico City are San Luis Potosí, Puebla, León and Orizaba.[25] Census data from 1990 suggests that other median-size cities in central Mexico, including Cuernavaca, Pachuca and Querétaro (which were not included in the 1987 ENMAU), have also been important destinations for out-migrants from Mexico City.

Selectivity in out-migration flows from Mexico City to median-size cities is apparent. The 1987 ENMAU further suggests that individuals with higher educational and occupational profiles may constitute a significant proportion of recent out-migrants from Mexico City.[26] Data from the 1990 census confirms that almost half (46 per cent) of all out-migrants from Mexico City had higher education levels (10 years or more), and over one-third were engaged in higher occupations (for example, technicians, educators, artists, officials, directors and office workers), compared to figures of 32 per cent and 22 per cent respectively for the population of the city as a whole.

Correspondingly, one-third to one-half of in-migrants from Mexico City to the states within which the median-size cities of Pachuca, Cuernavaca, Querétaro and San Luis Potosí are located had higher educational and occupational levels. Since the middle-class population of Mexico City in 1990 represented only one-third of the total population,[27] but 40 per cent of all out-migrants, their rate of out-migration appears to be higher than that of other socioeconomic groups. The out-migration of middle-class individuals from Mexico City to median-size cities may represent a distinct and important component of internal migration flows in recent years.

The role played by explicit population and industrial redistribution policies in stimulating this and other out-migration flows from Mexico City has been minimal.[28] Rather, the diverse economic and industrial changes discussed above have probably played the major role in stimulating out-migration flows, particularly among middle-class families who tend to have greater job flexibility and more resources. The city's deteriorating physical and social environment – that is, ecological determinants – may also play an important role in the out-migration response of middle-class individuals.

The 1987 ENMAU found that roughly one-third (32 per cent of all residents interviewed in Mexico City indicated that they would like to leave, while in other major cities in the country less than 10 per cent wanted to do so.[29] In addition, in contrast to other cities where economic motives dominated, the main reasons given for wanting to leave Mexico City related to the city's physical and social environment.[30] One-quarter to one-half of potential out-migrants from Mexico City cited environmental factors related to stress (55 per cent), pollution (41 per cent) and violence (25 per cent) as reasons for wanting to leave. The 1987 ENMAU also indicated that the proportion of the population wishing to out-migrate was higher among middle-class individuals (42 per cent) than for any other socioeconomic group. Among these potential middle-class out-migrants, one-third to one-half indicated wanting to leave because of environmental factors.

DISCUSSION

The long-term implications of the above industrial, ecological and demographic trends for the future of Mexico City are not clear. The current deconcentration of industry and population out of Mexico City may signal a move towards a more balanced model of urban development in which population and industry are spread out more evenly among smaller and median-size cities. On the other hand, in the absence of better regulation the deconcentration of industry and population may only lead to the replication of unsustainable patterns of development and negative environmental conditions in other cities. In this context, the demographic and industrial growth of smaller cities in the country's central region may be viewed as part of a process of 'megalopolization', for which Mexico City remains the core.[31]

Within Latin America, great attention has traditionally focused on rural–urban migration flows resulting from the region's historic pattern of centralized development.[32] Recent trends in Mexico City, however, suggest the need to recognize the new character and determinants of internal migration flows that may be occurring in Latin America, including the ecological determinants of these flows.

Previous empirical studies and the development of conceptual frameworks that consider the ecological determinants of migration or the role which the environment plays in determining population movement are limited. In recent years the concept of environmental refugees has brought more attention to the ecological determinants of migration. Astrid Surhke defines two types of ecologically determined migration, distinguishing between *environmental refugees*, whose movement is stimulated by environmental conditions but is involuntary, and *environmental migrants*, whose movement is also stimulated by environmental factors, but is voluntary.[33] Environmental refugees move because environmental conditions leave them no alternative, while environmental migrants choose to 'migrate before the situation becomes so bad that they have no other alternative'.[34] Anthony Richmond has further suggested that a continuum exists between environmental refugees and environmental migrants, and that ecologically-determined migration may be more or less reactive or proactive, respectively.[35]

Given their greater resources and job flexibility, middle-class individuals and families may choose out-migration from an array of options that might also include moving to the suburbs or having a weekend house outside of the city. Middle-class out-migrants from Mexico City are, therefore, probably more proactive environmental migrants rather than reactive environmental refugees. Examples of the latter, by comparison, might be victims of drought, famine or technological disasters.

The consequences of the increased out-migration of middle-class individuals from Mexico City are ambiguous. The urban middle-class in Mexico has traditionally been characterized by 'political alertness',[36] and may represent a key political group in terms of environmental activism. Increased out-migration by this group may therefore represent a loss of political power for improving environmental conditions in the city. At the same time, it may contribute to the more balanced distribution of the country's population, and middle-class migrants who leave Mexico City for ecological reasons may bring greater environmental awareness and concern to their new destinations. Alternatively, since the middle class is generally associated with higher consumption levels (for example, greater vehicle use), they may simply replicate negative environmental conditions in their new places of residence.

In any case, the middle-class still represents only a minority (about one-third) of Mexico City's current population. The responses and strategies being adopted by the majority of the population, who are of lower socioeconomic status, in the face of the city's deteriorating environmental conditions remains to be investigated. Out-migration may not be a realistic option for most inhabitants, since they may lack the occupational flexibility and the resources to move.

In general little existing research has addressed the impact of social class on population and environment relationships.[37] A greater understanding of the influence of socioeconomic status on population responses to and impacts on the environment are particularly important in the urban areas of Latin America, where social inequality is pronounced.

Since 1990, Mexico has experienced a series of profound political and economic changes, triggered by the ratification of the North American Free Trade Agreement (NAFTA), civil unrest in the southern state of Chiapas, and a major devaluation of the country's currency. The impact of these events on the economic, demographic and ecological trends discussed above remains to be investigated. An important question that arises is whether the industrial contraction, slower urban population growth, increased out-migration, and deteriorating environmental conditions observed in Mexico City during the late 1980s has continued into the early 1990s.

Preliminary research on the effects of the most recent economic crisis on internal migration in Mexico suggests that impacts may not be easily predicted. Despite the alarmist cries of American politicians, the most recent Mexican economic crisis appears to have triggered a sharp decrease in migration towards US–Mexican border states and into the US.[38] This may point to a general slowing in internal migration flows throughout the country. In this case, rapid rates of in-migration to Mexico City have probably not resumed, while out-migration flows from the city may have abated.

Data from both the next census and *ad hoc* surveys will eventually allow the magnitude and direction of the most recent internal migration flows in and out of Mexico City to be analysed. The challenge lies in understanding the array of factors driving the increasingly complex migration flows affecting Mexico City, as well as other Latin American cities. Given the deteriorating environmental conditions in many of these cities, the ecological determinants of these flows should be considered alongside the traditional economic determinants of migration. Continuing research on the ecological determinants of migration will contribute to a greater understanding of population and environment relationships and of internal migration as well.

Notes

1. United Nations, *World Urbanization Prospects* (New York: United Nations, 1990); and A. Lattes, 'Population Distribution and Development in Latin America', paper presented at the United Nations Expert Group Meeting on Population Distribution and Migration, Santa Cruz, Bolivia, 18–22 January 1992.
2. G. Garza and S. Rivera, 'Desarrollo Económico y Expansión Urbana en México', *Monografias Censales de México, 1990* (Aguascalientes: INEGI-IISUNAM, 1994). Because of possible overestimation of the population of Mexico City in the 1980 census and underestimation in the 1990 census the real decline in urban growth may have been somewhat more or somewhat less than published statistics indicate. See R. Corona, 'Confiabilidad de los Resultados Preliminares del XI Censo de Población y Vivienda de 1990', *Estudios Demográficos y Urbanos* 6 (1991): 33–68; Composortega, Sergio (1992) 'Evolución y tendencias demográficas de la zona Metropolitana de la Ciudad de México' in Consejo Nacional de Población (CONAPO), ed., La Zona Metropolitana de la Ciudad de México: Problemática actual y perspectivas demográficas y urbanas. CONAPO, Mexico City, 4–20.
3. See, for example, M. Negrete, 'La Migracíon a la Ciudad de México: Un Proceso Multifacético', *Estudios Demográficos Urbanos* 5 (1990): 641–53; L. González and M. Monterrubio, *Tendencias de la Dinámica y la Distribución de la Población, 1970–1992* (Mexico City: Consejo Nacional de Población (CONAPO), 1992); and R. Corona, 'Migración Permanente Interestatal e Internacional, 1950–1990', *Comericio Exterior* 43 (1993): 750–62.
4. G. Garza, 'Crisis del Sector Servicios de la Ciudad de México', paper presented at the Conference on Socio-Demographic Effects of the 1980s Economic Crisis in Mexico, The University of Texas at Austin, Population Research Center, The Mexican Center, Austin, Texas, 22–25 April 1992.
5. H. Izazola and C.M. Marquette, 'Migration in Response to the Urban Environment: Out-Migration by Middle Class Women and their Families from Mexico City after 1985', in *Population and Environment in Industrialized Regions*, eds A. Potrikowska and J. Clarke (Warsaw: Polish Institute of Geography and Spatial Analysis, 1995).
6. B. Roberts, 'Urbanization and the Environment', paper presented at the DAWN/Social Science Research Council (SSRC) Workshop on Population and Environment, Hacienda Cocoyoc, Morelos, Mexico, 28 January–1 February 1992.
7. Garza and Rivera, 'Expansión Urbana', note 2 above.
8. J. Gamboa de Buen, *Ciudad de México* (Mexico City: Fondo de Cultura Económica, 1994).
9. Garza and Rivera, 'Expansión Urbana', note 2 above.
10. M. Negrete, B. Graizbord and C. Ruiz, 'Población, Espacio y Medio Ambiente en la Zona Metropollitana de la Ciudad de México', *Programa de Estudios Avanzados en Desarrollo Sustentable y Medio Ambiente, El Colegio de México, Serie Documentos de Trabajo No. 2*, 1993.

11. See, for example, R. Lacey, 'La Calidad del Aire en el Valle de México', *Programa de Estudios Avanzados en Desarrollo Sustentable y Medio Ambiente, El Colegio de México, Serie Documentos de Trabajo no. 1,* 1993; and A. Herrera, 'Contaminación en Aire, Agua y Suelo en la Ciudad de México', in *Medio Ambiente y Desarrollo en México* Vol. 2, ed. E. Leff (México City: Centro de Investigaciones Interdisciplinarias en Humanidades del UNAM, 1990), 547–80.

12. Ozone is a secondary product created by the interaction of primary emissions of nitrogen oxide and hydrocarbon with sunlight.

13. E. Ezcurra, 'The Basin of Mexico', in *The Earth Transformed by Human Action,* eds B. Turner *et al.* (New York: Cambridge University Press, 1990), 577–88.

14. Herrera, 'Contaminación', note 11 above.

15. See, for example, H. Bravo, G.R. Ocotla, P. Sánchez and R. Torres, 'La Contaminación Atmosférica por Ozono en la Zona Metropolitana de la Ciudad de México', in *La Contaminación Atmosférica en México,* ed. I. Restrepo (Mexico City: Comisión Nacional de Derechos Humanos, 1992).

16. Gamboa de Buen, 'Ciudad', note 8 above.

17. See, for example, Lacey, 'La Calidad', *op. cit.* and R. Pèrez, 'Estadísticas de Energía y Contaminación Atmosférica en México. Hacia un Sistema de Estadísticas Ambientales', in *Población y Ambiente ¿ Nuevas Interrogantes a Viejos Problemas?* eds H. Izazola and S. Lerner (Mexico City: Sociedad Mexicana de Demografía, El Colegio de México, The Population Council, 1993), 105–30. Through information collected via the RAMA Network, a daily Metropolitan Air Quality Index (Indice Metropolitano de la Calidad del Aire-IMECA) has been calculated since 1986 and represents a combined index of emissions. The IMECA ranges from 0–500, with an index over 100 judged as prejudicial to human health. The IMECA index was 200 or more every six days in 1988, compared to every other day in 1991.

18. Negrete *et al.,* 'Población', note 10 above.

19. Although industrial and demographic concentration was most intense in Mexico City, similar processes occurred in Guadalajara, Monterey and Puebla from 1950 to 1970, also contributing to the country's metropolization. Together, Mexico City and these three cities accounted for over half of the total urban population by 1980, and formed 'the spinal cord of the country's national economy ... and the complementary process of urbanization'. G. Garza and S. Rivera, 'Desarrollo Económico y Distribución de la Población Urbana en México, 1960–1990', *Revista Mexicana de Sociología* LV (1993): 186.

20. Negrete *et al.,* 'Población', note 10 above.

21. See, for example, J. Finkelman, 'Medio Ambiente y Desarrollo en México', in *Medio Ambiente y Desarrollo en México,* vol. 2, ed. E. Leff (Mexico City: Centro de Investigaciones Interdisciplinarias en Humanidades, UNAM, 1990), 581–630; and C. Santos Burgoa and L. Rojas, 'Los Efectos de la Contaminación Atmosférica en la Salud', in *La Contaminación Atmosférica en México. Sus Casa y Efectos en la Salud,* ed. I. Restrepo (Mexico City: Comisión Nacional de Derechos Humanos, 1992).

22. M. Castillejos, 'La Contaminación Ambiental en México y sus Efectos en la Salud Humana', in *Servicios Urbanos, Gestión Local y Medio Ambiente*, eds M. Schteingart and L. d'Andrea (México: El Colegio de México, 1991), 187–204.

23. Consejo Nacional de Población (CONAPO), *Caracteristicas Principales de la Migración en las Grandes Ciudades del Pais: Resultados Preliminares de la Encuesta Nacional de Migración en Areas Urbanas (ENMAU)* (México City: CONAPO, 1987).

24. INEGI, *Programa de 100 Ciudades: Indicadores Sociodemográficos y Económicos* (México City: INEGI, 1993); Garza and Rivera, 'Expansión Urbana', note 2 above.

25. Negrete, 'La Migracíon', note 3 above.

26. R. Corona and R. Luque, 'Cambios Recientes en los Patrones Migratorios a la Zona Metropolitana de la Ciudad de México (ZMCM)', *Estudios Demográficos Urbanos* 7 (1992): 575–86.

27. INEGI, *Programa de 100 Ciudades*, note 26 above.

28. See, for example, Garza and Rivera, 'Expansión Urbana', *op. cit.* Roberts, 'Urbanization', *op. cit.* and C. Ruiz Chiappeto, 'Distribución de Población y Crisis Económica en los Años Ochenta: Dicotomía y Especulaciones', *Revista Mexicana de Sociología* 1 (1990): 185–204.

29. CONAPO, *Caracteristicas Principales*, note 23 above.

30. CONAPO, *Caracteristicas Principales*; Negrete, 'La Migracíon'; and Negrete *et al.*, 'Población', *op. cit.*

31. Garza and Rivera, 'Expansión Urbana', note 2 above.

32. See, for example, L. Arizpe, 'Relay Migration and the Survival of the Peasant Household', in *Why People Move*, ed. J. Balan (Paris: UNESCO, 1982); O. Oliveira, M. Pepin Lehalleur and V. Salles, *Grupos Domésticos y Reproducción Cotidiana* (México City: UNAM Coordinación Humanidades, El Colegio de México, and Grupo Editorial Miguel Angel Porrúa, S.A., 1988); Roberts, 'Urbanization', *op. cit.* and G. Hugo, 'Migration as a Survival Strategy: The Family Dimension of Migration', paper presented at the United Nations Expert Group Meeting on Population and Migration, Santa Cruz, Bolivia, 18–22 January 1993.

33. A. Suhrke, 'Pressure Points: Environmental Degradation, Migration and Conflict', *Environmental Change and Acute Conflict Project, Occasional Paper Series*, no. 3 (Toronto: International Security Studies Program, American Academy of Arts and Sciences, and Peace and Conflict Studies Program, University College, University of Toronto, Canada, 1993).

34. Suhrke, *ibid.*, 9.

35. A. Richmond, 'Environmental Refugees and Reactive Migration: A Human Dimension of Global Change', paper presented at the International Union for the Scientific Study of Population and Environment (IUSSP), General Meeting, Montreal, 24 August–1 September 1993.

36. S. Loaeza and C. Stern, 'Las Clases Medias en la Coyuntura Actual', in *Cuadernos del Centro de Estudios Sociológicos (CES)/El Colegio de México* (Mexico City: El Colegio de México/Centro Tepoztlán, 1990).

37. C. Marquette and R. Bilsborrow, *Population and Environment in Developing Countries: Literature Survey and Bibliography of Recommendations*

for Future Research, ESA/P/WP.123* (New York: United Nations, 1994).
38. J. Bustamante, 'El Tratado de Libre Comercio, Las Asimetrías Demográficas y Sociales', paper presented at the Plenary Session of the 5th Reunion of the Mexican Demographic Society (SOMEDE), Mexico City, 5–9 June 1995.

Reference

Legorreta, J. and A. Flores (1992) 'La Contaminación Atmosférica en el Valle de México', in I. Restrepo (ed.), *La Contaminación Atmosférica en México: Sus Causas y Efectos en la Salud* (Mexico City: Comisión Nacional de Derechos Humanos).

12 The Environmental Impacts of Refugee Settlement: A Case Study of an Agricultural Camp in Zambia

Véronique Lassailly-Jacob

REFUGEES AND THE ENVIRONMENT

Debates about the interconnections between environmental change and forced population displacement are relatively new, and little academic research has so far been produced on this topic. Meanwhile, the international community is increasingly concerned about environmental disruption which is recognized as both a cause and a consequence of population movements. It is a cause when:

> Growing numbers of people ... can no longer gain a secure livelihood in their homelands because of drought, soil erosion, desertification, deforestation and other environmental problems. In their desperation, these 'environmental refugees' as they are increasingly coming to be known ... feel they have no alternative but to seek sanctuary elsewhere, however hazardous the attempt.[1]

It has been estimated, that up to 10 million people in the world today are uprooted not by political turmoil, violence or persecution, but due to the degradation or destruction of their environment.[2] However, the concept of an 'environmental refugee' is still very controversial. JoAnn McGregor argues that '[t]he category "environmental refugee" confuses rather than clarifies the position of such forced migrants, since it lacks both a conceptual and a legal basis'.[3] Still labelled as economic migrants, many such forced migrants fall outside the categories protected by instruments of international refugee law: environmental decline is not yet recognized as a legitimate cause of refugee movements.

Environmental degradation can also result from the mass movement of people, both in the departure and the receiving areas. In the departure areas, damage may occur when farmers leave behind fields and pastures they have worked for centuries. In the mountainous rural areas of Afghanistan deserted by their inhabitants, highly developed irrigation systems have broken down, and the lack of maintenance of terraced land has led to erosion and the loss of fertile soil. Environmental degradation can also occur in receiving areas when large numbers of refugees concentrate in camps, generating heavy pressure on surrounding natural resources such as wood, water and arable land. This chapter will focus on the environmental changes resulting from the mass movement of people into receiving areas.

When Third World rural refugees arrive in their countries of first asylum, they may either spontaneously settle in border areas, or be channeled by host governments towards assigned locations – transit camps, care and maintenance camps, or agricultural settlements – where they are registered and assisted. Based on the model of land-settlement schemes, the first agricultural settlements for refugees in Africa were set up with assistance from the United Nations High Commissioner for Refugees (UNHCR) in the early 1960s. Designed to promote refugee self-sufficiency and local integration, these settlements were proposed as one of three durable solutions to the refugee problem (the other two solutions were related to repatriation and resettlement). The settlements have received considerable attention and financial support from the international community and from many local and international NGOs.

So far, the agro-ecological impacts on receiving areas of both scheme-settled and self-settled refugees has remained relatively understudied. Refugee assistance programmes have started paying attention to the environment only recently, since the energy crisis of 1973. Prior to this time, the impacts of the mass arrival of refugees were seen exclusively in terms of pressures on the economy and the infrastructure of the host area. The international community is now calling for increased attention to be paid to the ecological impacts of large concentrations of refugees. UNHCR assistance programmes for refugees have become increasingly aware of environmental problems.

The following analysis is based on an empirical study conducted by the author in an agricultural settlement in Zambia in 1993. The study's objectives were to assess the agro-ecological impacts of refu-

gees and to analyse the consequences of their presence on the livelihoods of host peoples. We argue that these agro-ecological impacts were an important factor straining refugee–host relationships,
but they were not the main source of tension. Rather, the interventions of aid organizations may have significantly affected the
quality of host–refugee relationships, and may in fact have jeop-
ardized some refugees' means of livelihood by disrupting crucial
relationships between them and their hosts.

THE SETTLEMENT OF MOZAMBICAN REFUGEES IN UKWIMI, ZAMBIA

Ukwimi, an agricultural settlement, was established in 1987 to house
Mozambican refugees. As a result of the civil war that had been
raging in Mozambique since 1976, an estimated 1.7 million
Mozambicans had sought refuge in neighbouring countries to escape intense fighting between the Mozambican Liberation Front
(FRELIMO), which held power, and the Mozambican National
Resistance (RENAMO). They found refuge in Malawi, the 'homelands' of South Africa, Zimbabwe, Tanzania, Zambia and Swaziland.
Mostly of rural origin, these refugees either spontaneously settled
in border areas, or were channeled by host governments towards
reception or transit camps (where emergency assistance was provided), towards care and maintenance camps (where basic facilities as well as access to vocational and income-generating activities
were provided) and towards agricultural settlements (where development assistance was given, mainly for farming) (see Figure 12.1).
These agricultural settlements were supposed to help the refugees
become self-sufficient and locally integrated.

In 1984–85, due to the growing number of RENAMO attacks in
Tete province, many refugees arrived in Zambia and spontaneously
settled in the border area. In 1987, RENAMO carried out cross-
border raids, attacking the villages and making the area a security
risk for both Zambians and Mozambicans. In addition, the increasing
population density was putting pressure on land, water and natural
resources and local services. The Zambian government decided to
regroup and transfer all of the spontaneously settled refugees to a
distant agricultural settlement called Ukwimi. In the following years,
new waves of refugees arrived in Ukwimi, often fleeing the war,
but also driven by drought and famine. The last refugees arrived

Figure 12.1   Mozambican refugees in neighbouring countries, 1993

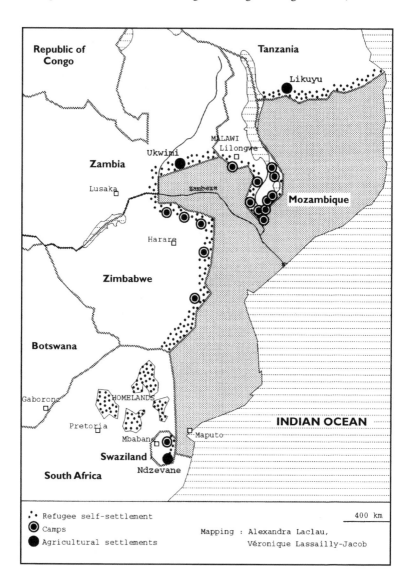

Figure 12.2   Mozambican refugees in Zambia

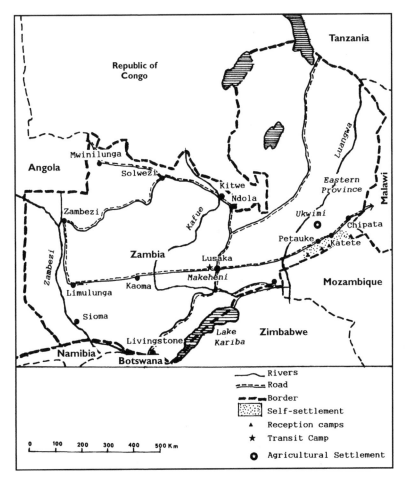

there in March 1993. A peace agreement signed in October 1992 between the Mozambican government and RENAMO concluded a war that had devastated the country, Since then, repatriation has continued at a slow pace. The refugees left Ukwimi at the end of 1994, leaving behind a well-planned and equipped site with cleared land and substantial infrastructure. Local people and new Zambian settlers now occupy the site.

Ukwimi was implemented under a tripartite agreement between UNHCR, the Zambian government, and the Lutheran World Fed-

Figure 12.3   Ukwimi, layout and management in 1993

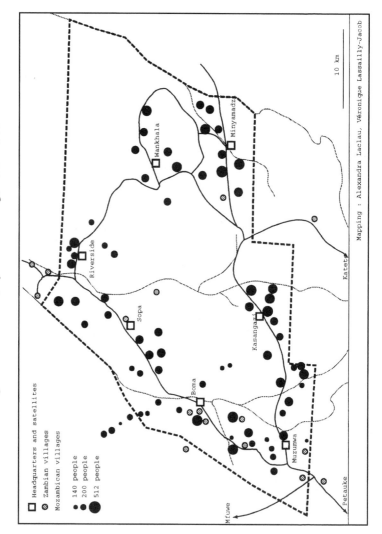

Mapping : Alexandra Laclau, Véronique Lassailly-Jacob

eration (LWF). At the time of research, the settlement was administered by the Commission for Refugees under the Zambian Ministry of Home Affairs and run by LWF (the lead organization) and Refugee Services of Zambia (RSZ), a local NGO. Several other implementing partners also operated in Ukwimi.

Ukwimi lies in an isolated location in the Eastern Province of Zambia, 500 kilometres (km) from Lusaka, 70 km north of Petauke, and more than 100 km from the Mozambican border (see Figure 12.2). It was established in hilly terrain on the edge of the Luangwa escarpment, at an altitude of 800 metres (m), where the vegetation is predominantly dry *miombo* and *mopane* woodland. The area has a dry tropical climate, with the annual average rainfall of 800 mm usually occurring between November and March; however, the region is prone to periods of inadequate rainfall that frequently threaten crop yields. Three streams water the settlement all year round.

This well-managed settlement had two headquarters, Ukwimi A and Ukwimi B, each supervising a number of satellites, four under Ukwimi A, and three under Ukwimi B. Each satellite in turn served 8 to 14 refugee *'villages'* and neighbouring Zambian villages (see Figure 12.3). The locations and sizes of the refugee villages were determined on the basis of hydrologic and soil surveys. Boreholes were drilled and equipped with hand pumps; and dry season water yields determined the size of the villages, which usually ranged from 200 to 300 people. Each newly-arrived family was allotted a homestead of 600 square metres, which they had to clear for their huts.

At the time of this research during the summer of 1993, there were about 29 000 inhabitants in the camp. They were divided into three separate communities inside the settlement boundaries: 25 600 were refugees settled in 73 *'villages'*; over 3000 were local Zambians in 9 villages and 5 hamlets; and about 500 staff people were concentrated in the two headquarters sites and five of the satellites.

The settlement is widely spread over 310 square kilometres, but the boundaries were never precisely defined, often leading to disputes over land rights. Within the three communities land was divided between *'pockets'* of land owned and farmed under customary law by the local Zambians, and government land that was allocated to refugees and staff. The government gave each Zambian village a demarcated territory that remained under the authority of the traditional chieftainship. Zambian farmers were no longer allowed to practice shifting cultivation or to resettle outside their new boundaries as they had done in the past. Land outside of these boundaries was declared government property and divided between refugees and the staff. Each refugee head of household was allocated a farm plot of two hectares (ha), which had to be cleared and cultivated; agricultural tools, fertilizers and seeds for soy beans, rice, maize and groundnuts were also provided. After two years the refugees were supposed to be self-sufficient in terms of their food requirements. The staff received arable plots around the headquarters and the satellites, but they usually asked the local people for additional land.

Refugees and their hosts also benefited from high-quality facilities, including primary schools, clinics, roads, clean water, markets, shops, grinding mills and warehouses. Returning from Ukwimi in 1990, R. Black *et al.* reported that 'Ukwimi is probably one of the best implemented agricultural settlement programmes for refugees in Africa, and has been widely promoted as a model for development assistance to refugees'.[4]

The local Zambians and the Mozambican refugees were related to each other as they both belonged to the Chewa and Nsenga ethnic groups. They spoke the same language and shared a common culture and set of traditions. In addition, they were dispersed over a large area in contrast to the overcrowded camps in Malawi and elsewhere, and all benefited from the provision of high-quality infrastructure and services, which had collapsed elsewhere in the Zambia. Nonetheless, both communities voiced complaints, and land disputes were increasing threatening the smooth implementation of the assistance programmes. Why did these two communities come into conflict, despite their many reasons to cooperate?

THE AGRO-ECOLOGICAL IMPACTS OF THE SETTLEMENTS

Since most refugees came from rural areas, they began to reproduce their traditional gathering activities after arriving in Ukwimi. Thus, both communities relied on the local environment to meet similar daily subsistence needs for energy, shelter, medicine, food and income. Both local people and refugees collected dead wood to use for fuel and charcoal, poles for building houses and granaries, bamboo for making doors and furniture, reeds for weaving mats, building bathing shelters, and fencing homesteads and gardens, grass for thatching houses, earth for plastering walls, and plants, tubers, tree bark and sap for medicinal purposes. Obviously, the heaviest demands for shelter resources and the resulting environmental degradation occurred at the time of the refugees' arrival, during the initial emergency phase.

Refugee households relied more heavily on the environment to fulfill needs for food and income than did locals, and refugees looked for bush food to supplement their diet. During their first two years in the settlement all refugees received the same *'food basket'*, provided every two weeks by the World Food Programme. Each household received a ration of 500 grams of maize, 20 grams of beans, 15 grams of sugar, 20 grams of vegetable oil, 10 grams of salt and 60 grams of soy beans per person per day. These food rations were inadequate, at least in terms of variety, to meet the refugees' overall nutritional needs. Ken Wilson (1992)[5] argues that this diet was deficient in calories and micro-nutrients, and, as such, it was unpalatable.

To supplement and diversify this monotonous diet, refugees sought to obtain items for *relish,* a side-dish, especially to accompany the *nshima* maize porridge the main course at the two daily meals. Relish, an important part of traditional meals, improves the nutritional quality and varies the diet. It can be made from edible leaves (mainly pumpkins and beans), vegetables, tubers, mushrooms, fruits, meat, fish, honey, insects, rodents, snails, caterpillars and other items that come mainly from the natural environment, but also occasionally from cultivated crops. In order to obtain relish and additional food, refugees initiated diverse *'diet supplementation strategies'.*[6] These strategies can be categorized according to the origin of the supplement in either communal lands, the lands of local Zambians, or refugees' farmland.

Under the authority of the traditional chiefs, the resources on communal land are entirely open to all. Refugees could obtain additional food by depleting natural resources that Zambian villages owned and controlled collectively. This depletion arose from two types of activity. First, refugees gathered bush foods, trapped game animals and birds, and fished. In addition to adding seasonal diversity to meals, such *'bush foods'* are also consumed during food shortages. During the rainy season, women and children gather wild fruits and leaves, tubers, mushrooms, insects and caterpillars. At the time of this research, leaves and mushrooms were laid out to dry on thatched roofs or hung inside houses. Second, refugees set fires during the dry season to catch rodents, rats, rabbits and mice.

Refugees also obtained food from local Zambians. Upon arriving at Ukwimi, they found a rural host community that could provide them with many of the food items that were lacking. Local

Zambians were growing pumpkins, okra, sweet potatoes, cassava, mangoes, papayas, guavas and bananas, and rearing chickens, ducks, goats, pigs and pigeons. In order to obtain these additional food products, as well as local seeds (for maize, pumpkin, sorghum or sweet potatoes), refugees bartered using items from their rations, purchased items, or worked for the Zambians (digging, weeding or harvesting on their farms).

Finally, refugees obtained food from their 2 ha farms, planting maize, groundnuts, beans and, sometimes, rice, sunflower or tobacco. Seeds, fertilizer and other agricultural inputs were provided free for the first two years. After this, refugees were supposed to be able to meet their food requirements and earn an income from their farms. However, some household had been allocated plots with poor soil in areas of low agricultural potential, and the system of continual cultivation on these fragile areas had already exhausted the soil. Most of these *'unfortunate'* farmers, as well as other enterprising refugees, sought to extend their farms by seeking permission from local people to cultivate additional plots. Refugees located near the valley areas decided to grow rice and to keep dry season vegetable gardens in *dambos*, valley areas that flood during the wet season. Local people allotted the refugees farm plots in the upland and *dambo* areas. Those refugees who cultivated a garden in *dambo* lands were assisted by LWF with free vegetable seeds (for cabbage, rape, tomato and onion) and free fertilizer for two years. As a result, refugees developed dry season vegetable gardens in valley areas owned but not used by local Zambians.

Refugees also used natural resources as raw materials for income-generating activities, especially since they were not allowed to leave the settlement without permission. This put added pressure on the local environment. While women specialized in making clay-pots and brewing beer from maize and sorghum, men concentrated on weaving baskets and mats, carving wood and doing carpentry, constructing houses and granaries, and selling timber and charcoal. As a result the area was depleted of wild resources and game. In addition bush fires partly deforested the area, and the land was being farmed much more heavily than it had been prior to the arrival of the refugees. Through interviews and field observations, the consequences of this environmental degradation on the local population could be assessed.

IMPACTS ON LOCAL LIVELIHOODS AND
REFUGEE–HOST RELATIONS

Due to these intensive gathering activities, natural resources were depleted in the more populated areas of the settlement. In a report on her fieldwork in Ukwimi, Spitteler explained that 'Food derived from woodland habitat, particularly important in drought prone areas such as Ukwimi where woodland food is a "buffer" against famine, was becoming increasingly difficult to find'.[7] Many Zambian respondents talked about the difficulty of finding bamboo, reeds, wild animals and even caterpillars.

Since most local Zambians used bush food for both home consumption and sale, this depletion affected their livelihoods. During her fieldwork in Ukwimi, Sullivan reported that 'Resource competition and the resultant shortage of gathered products has affected Zambian incomes through reducing the availability of these products for sale and has generally had negative impact on diet and general well-being'.[8] Local Zambians who wished to gather bush food had to travel far outside the settlement. Moreover, while gathering wild resources was traditionally a female-dominated activity (except for collecting honey and hunting), it was becoming an increasingly male-dominated activity because of the long walking distances involved. As a result, local women had lost this important traditional source of income.

Changes in the patterns of setting bush fires also had an impact. These fires were traditionally set late in the dry season, in August and September, in order to clear the land for the new farming season. By this time all crops had been harvested and stored, and wild grass had been collected for thatching huts or fencing gardens. The refugees, however, especially the children,[9] set bush fires much earlier, in June or July, in order to catch rodents. These early bush fires deprived the local population of thatching grass and threatened the maize crop which was left standing until transportation was available to carry it to granaries. Refugees, on the other hand, stored their maize early because LWF had provided each refugee village with two oxen and a cart. After carrying in the crops produced by the refugees, the carts and oxen could then be hired by Zambian people and staff.

The depletion of wild resources and the premature setting of bush fires by refugees caused the local population to become bitter and resentful of the refugees. Whole trees were being cut down to collect edible caterpillars, or to dig up the roots and remove the bark. Local Zambians accused the refugees of improperly managing the environment; they considered them outsiders who did not properly care for an environment that did not belong to them. One Zambian respondent declared that:

> They chop down the trees we worship, they chop down very old trees along the streams in order to extend their gardens when these trees helped prevent evaporation with their shade. They collect all the bamboo and reeds growing at one place, they use chemicals to kill fish, and they wash their clothes in the stream, polluting the water we drink; they also light bush fires too early, when our maize is still in the fields.

This depletion of wild resources has had an especially negative effect on poor Zambians, who rely most heavily on bush resources for their livelihood.

The environment also suffered from the clearance of bush and the depletion of game. Prior to the establishment of Ukwimi, the site was covered by large areas of bush in which game animals dwelled, leading to the spread of tsetse fly in the area.[10] Because of hunting by the refugees, local Zambians could no longer obtain much appreciated 'bush meat'. At the same time, bush clearing and depletion of game resulted in an enormous reduction in the tsetse fly threat in the area, and meant that fewer wild animals threatened local crops. As a consequence, local Zambians were able to raise more livestock and start farming *dambos* to grow vegetables for home consumption and marketing.

Prior to the refugee settlement, only a few Zambian households farmed small dry-season gardens in *dambo* lands, in which they grew bananas, sugar cane and some vegetables. *Dambo* lands had not been thoroughly exploited for several reasons. First, part of the settlement was a game reserve, and the abundant wildlife in the area, including elephants, monkeys and bush pigs, damaged the gardens which could not be adequately guarded due to the absence of most adult men who were working in cities. Secondly, transportation problems made it difficult and expensive to obtain seed, fertilizer and insecticides. Finally, there was not much need to cultivate vegetables because bush food was so plentiful.

Zambian headmen, who are the village chairmen and local leaders, benefited most from the development of *dambo* cultivation because they were in charge of distributing farmland. They and their relatives opened new gardens or expanded previous gardens. They also offered land to refugees and staff people, and in return they received many presents such as vegetables, seeds or fertilizer as tokens of appreciation. In contrast, ordinary local households had less access to *dambo* land, since much of was already occupied by refugees, staff people, Zambian headmen and wealthy Zambians. Moreover, even when ordinary local households did have access to *dambo* land, they could not obtain vegetable seeds or fertilizer. They resented having to buy vegetables grown on their own land by refugees who received a great deal of assistance. Disputes between refugees and Zambians over *dambo* land increased, especially in the northern part of the settlement where the density of the refugee population was very high. In this area, more than 100 refugees had been evicted by the locals from the *dambo* gardens they had been cultivating.

The environmental impacts of the refugees were therefore perceived both positively and negatively by local Zambians, depending on their gender and social status. As suggested by Chambers,[11] there were winners and losers among the host population due to the refugees' environmental impacts.

NGOs AND REFUGEE–HOST RELATIONS

In the case of self-settlement, only two partners interact: refugees and their hosts. In the case of scheme-settlements, however, another important player has to be considered: the aid agencies or NGOs. Aid agencies play a major role in planning and designing the site, reallocating resources and distributing assistance. Interactions and relationships between scheme-settled refugees and their hosts cannot be understood without studying the interventions of aid agencies as well. What policies did NGOs in Ukwimi implement?

First, the area selected for the settlement was relatively isolated, and conditions were marginal in terms of the quality of its climate, soils and water supply; 'Ukwimi was selected for a refugee settlement not on grounds of ecological or economic suitability, but because it was nearly uninhabited'.[12] The settlement was spread over 31 000 ha, but about half of this area was uncultivated and uninhabited because of infertile soils, steep slopes, or difficult access to groundwater. Selecting a site with such low agro-ecological potential for the settlement of a large refugee population that would rely heavily on its environment for survival meant that the sustainability of the settlement would be at risk.

Secondly, too many refugee villages were located in certain areas. According to the 1992 Ukwimi annual report by LWF, over 14 000 ha were occupied by hills, rivers and other uncultivatable areas. This meant that a population of about 30 000 (including Zambian villagers and staff) was concentrated on 15 000 ha, a density of 200 persons/km^2. Implementing agencies did not pay enough attention to the sustainable carrying capacity of the site or to the potential impacts on the local environmental of such a high rural population concentration. They underestimated the pressure on natural resources that would inevitably lead to competition and conflict with the host population.

Thirdly, the agencies initiated a programme of environmental rehabilitation that did not address the needs of the Zambian community. In 1988, the lead implementing agencies started a programme of 'environmental rehabilitation' in order to overcome the deforestation that was affecting an estimated 5000 ha. The agencies decided to plant fast-growing exotic species (Eucalyptus, Gmelina and Cassia) along roads and on 83 ha of woodlands. In addition they prohibited access to approximately 2500 ha of forest, which were designated a 'reserve'. Zambians villagers were asked to work on the woodlots as part of a Food for Work programme, but they refused arguing that these trees would not benefit them, but the administration. They argued that the selected species would only provide poles and poor quality firewood; they would have preferred indigenous species that have more uses.

Finally the NGOs intervened on behalf of the refugees. The policy of the aid agencies was that the refugees should be self-sufficient after two years, so food and agricultural assistance was to be stopped at that time. In practice, however, food assistance was cut, but not assistance with agricultural inputs and other support. Refugees continued to receive subsidized agricultural inputs, such as seeds and fertilizers provided on a loan basis. In contrast, local Zambians did not receive any help. During 1993 fieldwork, one Zambian respondent complained that:

> We understand that refugees had nothing when they arrived here and they should be helped and assisted, and we agreed to give them land to feed them. But we see that now many refugees make much profit from our land. They are now growing tobacco, sunflowers and vegetables, and they are provided with seeds and fertilizers when we get nothing.

In addition, the refugees continued having free access to medical care (in the hospital), and schools (free uniforms and exemption from examination and boarding fees in secondary schools), whereas local Zambians had to pay for these services.

Finally, the programme for dealing with the severe drought of 1991–92, which affected the two communities equally, was not the same for refugees and local people. This severe drought was not due to inadequate rainfall (it rained 734 mm in Ukwimi), but to lack of rain in the critical months of January and February. When rains did fall in March, it was too late to save the crops. Refugees received emergency relief food, such as maize and beans, and they were provided with groundnut seeds and fertilizers through loans, and with drought-resistant seeds for sorghum and cassava. They also received other significant donations. In contrast, local Zambians received only maize, cooking oil and beans under the condition that they participate in Food for Work programmes. Some poorer Zambians actually went to work on refugees' fields to obtain groundnut seeds for the next season.

NGOs viewed and administered refugees and Zambians as two different groups. Although some refugees were very successful as farmers or woodcarvers and earned more than local Zambians, they continued to receive assistance because of their refugee status. A number of refugees invested their profits in shops and grinding mills – most private shops and mills in the settlement were owned by refugees. In addition, many refugees had bicycles and radios while few Zambians could afford them. This situation also led to animosity and resentment within the host community.

CONCLUSION

While it is widely acknowledged that rural societies in a subsistence economy rely heavily on their environment to satisfy a variety of basic needs, it is too often forgotten that rural refugees will likewise be heavily dependent on the natural resources of the reception area for a significant part of their livelihood. The ecological impact resulting from large refugee concentrations have too frequently been overlooked. In Ukwimi, refugees have undoubtedly had a major environmental impact due to their needs for shelter, food and income. This has caused economic hardships for poor Zambians and women.

However, in Ukwimi, resource pressure and environmental damage occurred in the most populated parts, and mainly during the early stages of settlement. Furthermore, the host community perceived the environmental impact of the refugee settlement in two ways. There were some positive impacts, such as the eradication of tsetse fly and the disappearance of destructive wild animals. But negative impacts arose from the depletion of wild resources and destructive early bush fires. But this damage was only part of the reason for the antagonism that arose between the two communities leading to the exclusion of many refugees from cultivating gardens on *dambo* lands. Aid organizations also bear part of the responsibility for this growing animosity. Their decisions to select a marginal site with respect to both its location and its agro-ecological potential, to concentrate the refugee population in some areas of the settlement, to undertake a reforestation programme that did not respond to the needs of local Zambians, and, above all, to distribute post-emergency assistance only to the refugee community, all contributed to the tensions that developed between the two communities.

The literature on these issues places too much emphasis on environmental arguments as the major causes of the strained relations between refugees and their local hosts, while other factors are overlooked. In Ukwimi, the discriminatory interventions by aid organizations were a more significant cause of this problem. These interventions went so far as to jeopardize the livelihoods of both local Zambians and refugees by disrupting the crucial relationship between the two groups. Aid organizations can play a major role in determining the quality of refugee–host relations and the potential for successful integration of these communities.

Notes

1. N. Myers and J. Kent, 'Environmental Exodus: An Emergent Crisis in the Global Arena', *Climate Institute* (1995): 1.
2. J. Jacobson, *Environmental Refugees: A Yardstick* of *Habitability* (Washington, DC: Worldwatch Institute, 1988).
3. J. McGregor, 'Refugees and the Environment', in *Geography and Refugees: Patterns and Process of Change* (London: Belhaven Press, 1993), 157.
4. R. Black, T. Mabwe, F. Shumba and K. Wilson, 'Ukwimi Refugee Settlement: Livelihood and Settlement Planning', report for UNHCR and the Government of Zambia (London: King's College, 1990), 1.

5. K. Wilson, 'Enhancing Refugees' Own Food Acquisition Strategies', *Journal of Refugee Studies* 5 (1992): 226–46.
6. K. Wilson, D. Cammack and F. Shumba, 'Food Provisioning Amongst Mozambican Refugees in Malawi: A Study of Aid, Livelihood and Development', report prepared for the World Food Programme, Oxford University, 1989.
7. M. Spitteler, 'Balancing Woodland Resource Use Needs With Environmental Needs: A Case Study of Ukwimi Refugee Settlement, Zambia', fieldwork report (Oxford: Refugee Studies Programme, 1993), 2.
8. S. Sullivan, 'Utilisation of and Ecological Impact on Wild Resources by Mozambican Refugees and Local Zambians at Ukwimi Refugee Settlement, Zambia', fieldwork report (London: University College, 1992), 20.
9. The growth rate among the refugee population was very high; the crude birth rate reached 4.5 per cent in 1991. See Lutheran World Federation, *Annual Reports*, 1991.
10. L. Vail, 'Ecology and History: The Example of Eastern Zambia', *Journal of Southern African Studies* 3 (1977): 139.
11. See the following: R. Chambers, 'Hidden Losers? The Impact of Rural Refugees and Refugee Programs on Poorer Hosts', *International Mitigation Review* 20 (1986): 245–65; and R. Chambers, 'Rural Refugees in Africa: What the Eye Does Not See', *Disasters* 3 (1979): 381–92.
12. Black *et al.*, 8, note 4 above.

References

Black, R. (1994) 'Environmental Change in Refugee-Affected Areas of the Third World: The Role of Policy and Research', *Disasters* 18: 107–16.
Black, R. (1994) 'Forced Migration and Environmental Change: The Impact of Refugees on Host Environments', *Journal of Environmental Management* 42: 261–77.
Black, R. (1993) 'Refugees and Environmental Change: Global Issues', Report prepared for ODA Population and Environment Research Programme (Bradford, UK: University of Bradford).
Black, R. and T. Mabwe. (1992) 'Planning for Refugees in Zambia: The Settlement Approach to Food Self-Sufficiency', *Third World Planning Review* 14: 1–20.
International Organization for Migration and the Refugee Policy Group (1992) 'Migration and the Environment'.
Kibreab, G. (1989) 'Local Settlements in Africa: A Misconceived Option?' *Journal of Refugee Studies* 2: 468–90.
Lassailly-Jacob, V. (1993) 'Refugee–Host Interactions: A Field Report from the Ukwimi Mozambican Refugee Settlement, Zambia', *Refuge, Canada's Periodical on Refugees* 13: 24–7.
Lassailly-Jacob, V. and M. Zmolek (1992) 'Environmental Refugees', *Refuge, Canada's Periodical on Refugees* 12: 1–4.
Lutheran World Federation, *Annual Report*, 1990, 1992, 1993.

13 The Struggle over Transboundary Freshwater Resources: Social and Economic Conflict in the Appropriation and Use of Water along the US–Mexico Border[1]

Roberto A. Sánchez-Rodriguez

Recent international attention regarding water resources has focused on their scarcity and uneven distribution in several parts of the world. Some studies suggest that as many as 80 countries, representing 40 per cent of the world's population, already suffer from serious water shortages in some regions or at some times during the year.[2] Water is a fundamental resource to all environmental and societal processes, and disputes over transboundary water resources are increasingly leading to international conflict in several regions of the world. However, efforts to create a general (customary) international law and a forum to solve international water disputes – most notably by the International Law Commission sponsored by the United Nations, the Institut de Droit International, the Second International Water Tribunal, and the International Law Association – are still limited. Most agreements concerning shared water resources are bilateral and relate to specific rivers that form or cross boundaries, or to lakes that straddle borders. These controversies are best resolved when the bilateral relations between the states concerned are otherwise good, when it is in their mutual interest to reach agreement, or when the asymmetry between the states involved is clearly defined and the more powerful state decides to end the dispute.[3]

In contrast to the attention devoted to international conflicts over water, much less attention has been focused on the profound social and economic consequences arising from imbalanced distribution and access to water within states, which can also contribute to international conflicts. Struggles occur especially between those who depend on the economic functions of water as a key resource for generating economic growth, and those who rely on its social function as an indispensable resource for maintaining human life and community well-being. Although this struggle exists within all societies, its extent and consequences differ dramatically according to a country's level of development. The wider socioeconomic gaps that typically exist in developing countries are likely to heighten the polarization between different groups of water users. The social consequences of unequal access to water can be particularly significant for weaker social groups, affecting their health, income and standard of living.

This chapter presents an overview of the transboundary water resources disputes between Mexico and the US, which share the Colorado River and the Rio Grande/Rio Bravo, as well as groundwater resources. Despite a long history of water conflicts between these two countries and their relative success in resolving them in the past, the management of transboundary water resources is reaching a critical stage. These resources are already overexploited and polluted. In combination with the sharp increase in water demand in recent decades driven by economic advance, population growth and the emergence of new water uses, the legal and institutional framework controlling the use of these resources is now being pushed to its limits. However, the unique feature of this transboundary situation is that rather than leading to the reemergence of an international conflict, the increasing pressure on scarce transboundary water resources is apparently leading to the evolution of an informal transboundary water market.

The leading force behind the formation of a transboundary water market is the response of economic actors to water shortages and to the disfunctionalities of the existing legal and institutional framework. This framework, negotiated decades ago, was designed as a response to the needs of a regional economy that was based primarily on agricultural activities. But the shape of the regional economy in the southwestern US and in northern Mexico has shifted dramatically during the last few decades as the shares of industry, trade and services have increased. In addition, these economic changes have also led to rapid urbanization; the vast majority of the population in the border area is now concentrated in urban areas, which will be the site of growing demands for water.

The evolution of a transboundary water market is also linked to a variety of other factors, ranging from the impact of climatic variability on the availability of water, to the rise of new water uses that were not considered when the existing legal and institutional framework was developed, including environmental uses and Native Americans' claims. Competition for water will continue to escalate in the near future, as previously unutilized water rights in the Upper Colorado, the Rio Grande/Rio Bravo, and smaller basins along the US–Mexico border are increasingly being exploited. Implementation of the North American Free Trade Agreement (NAFTA) could also affect water demands, with potentially severe consequences for weaker water users, thus further facilitating the growth of the transboundary water market.

US–MEXICO WATER RESOURCES ISSUES AND CONFLICTS

Demands for Colorado River Water

Legal entitlements to water in the Colorado River basin were negotiated in 1922 by the seven US states in the watershed (Colorado, Utah, Wyoming, Nevada, Arizona, New Mexico, and California),[4] but the 1922 Colorado River Compact was later modified by the US–Mexico Water Treaty of 1944. However, this appropriation of water in the Colorado River Basin was negotiated based on an overestimation of the amount of water available: the actual average annual flow in the river is about 19 billion cubic metres (m^3), rather than the 22 billion m^3 level used in negotiation.[5] Nevertheless, subsequent laws and decrees have been based on the original apportionments agreed to in the 1922 Compact.

This 1944 treaty also established the International Boundary and Water Commission (IBWC) to administer its implementation. The commission members collect and exchange data, estimate demands and plan and investigate projects, while maintaining the facilities necessary under the treaty, and mediating disputes arising in the border area. In 1979, the IBWC signed Minute 261, which substantially expands its jurisdiction to address a range of water-related pollution problems.[6]

The salinity crisis of 1961–73 in the Colorado basin was a critical test of the ability of these two countries to resolve water disputes. The dispute arose after the Wellton Mohawk Irrigation District in Arizona began draining saline groundwater into the Colorado River and charging it against Mexico's 1944 Treaty allotment, with immediate and severe impacts on agriculture in the Mexicali Valley, one of Mexico's most productive agricultural regions. The response of Mexican officials was a crash programme of groundwater development along the Arizona border; they argued that the depletion of common groundwater resources was compensation for the injuries resulting from the increased salinity in the river.

The salinity dispute was eventually settled in 1973 by Minute 242 of the IBWC, which assured Mexico the same water quality as that delivered in the US based on an annual average, placed limits on groundwater withdrawals within five miles of the border, and committed each nation to consult with the other about any plans for future groundwater development in the region. Mexico's problem has not been completely resolved, however, in part because the low salinity levels in the Colorado during the winter and spring allow Arizona farmers to increase the salt concentrations during the rest of the year, causing damage to Mexican agriculture while still complying with the annual average established by Minute 242.

In the past decade a new conflict has arisen in the basin over the lining of the All American and Coachella Canals in Imperial Valley. The Mexicali Valley sits atop a large aquifer that extends north to the Imperial Valley, and Mexico has exploited this aquifer to support agriculture since 1955 through a network of more than 700 wells. Lining the canals – including 48 kilometres (km) of the All-American Canal, as well as 61 km of the Coachella branch – will sharply reduce the seepage that recharges the aquifer, and thus significantly reduce water availability in the Mexicali Valley, affecting at least 121 wells and more than 13 355 hectares (ha) of farmland, with damages estimated at more than \$80 million per year.[7] No agreement has been reached between Mexico and the US regarding this conflict.

Growing urban demands for Colorado River water in both the US and Mexico are also a source of concern. The Metropolitan Water District of southern California (MWD) now receives 678 million m^3 per year from the Colorado (down from 1500 million m^3 before the Central Arizona Project (CAP) was completed), but the MWD expects shortages in deliveries to its customers of at least 247 million m^3 per year by the year 2000 if it cannot secure additional water sources. At the same time that water availability is declining, the population served by the MWD, which includes the metropolitan areas of Los Angeles, Orange County and San Diego, is expected to grow from 15 million in 1980, to 23 million in 2010. Urban water demand in California is expected to increase by more than 30 per cent in 20 years, and meeting this need is among the most critical water problems facing the state.[8] The situation is similar in Baja California, Mexico, where population is expected to grow from 1.2 million in 1980 to more than 3 million in 2010.

The Rio Grande/Rio Bravo Basin

The other major source of surface water along the US–Mexico border is the Rio Grande/Rio Bravo, which extends more than 2000 km along the border between Texas and the Mexican states of Chihuahua, Coahuila, Nuevo Leon, and Tamaulipas. In the 1906 Convention for the Equitable Division of the Waters of the Rio Grande for Irrigation Purposes, Mexico agreed to the development of the Elephant Butte Dam in New Mexico, and the US committed itself to providing Mexico with 74 million m^3 of river water for Ciudad Juárez and the surrounding area south of the border. The 1944 US–Mexico Water Treaty established cooperative regulation of the Rio Grande/Rio Bravo from Fort Quitman to the Gulf of Mexico, provided for two international reservoirs – Amistad and Falcon – and set explicit guidelines governing diversions from the river and its tributaries.

The US typically takes more water from the Rio Grande/Rio Bravo than Mexico: 2700 million m^3 per year on average for the US, compare to 1330 million m^3 for Mexico, out of an average annual inflow of 4305 million m^3 (2505 million m^3 from Mexico, and 1800 million m^3 from the US). The Texas Water Development Board (TWDB) projects that by the year 2040, the Rio Grande/Rio Bravo and its associated tributaries and aquifers will fall short of being able to meet all of the demands placed on it by water users in the US by about 338 million m^3 per year. The bulk of this shortage – 211 million m^3 – is expected to fall on agricultural users.[9] The TWDB also projects that between 2000 and 2040, aquifer supplies will fall by 0.8 per cent annually for the basin as a whole.

The four largest metropolitan areas in the Rio Grande/Rio Bravo

Basin – El Paso/Ciudad Juárez, Laredo/Nuevo Laredo, McAllen/ Reynosa, and Brownsville/Matamoros – present particular problems for future water planning. Population pressures are projected to create water shortages in every area by the year 2040, although conservation measures could help keep demand within available supply limits, except in El Paso/Ciudad Juárez, which already depends heavily on groundwater. The TWDB forecasts a significant decline in total water supplies for El Paso County between 2020 and 2030, and similar shortages are also expected in Ciudad Juárez before 2030.

Water quality is also a problem in the Rio Grande/Rio Bravo, especially due to the discharge of untreated waste by the cities along the border, especially along the Mexican side, although the US side is not blameless. Non-point source pollution and the release of hazardous wastes into the water are also problems in several parts of the river.

Groundwater Resources

Groundwater provides over 60 per cent of all water used in Texas and Arizona, and six of the 17 border municipalities rely solely on groundwater, while another four are partially dependent on it. Agricultural communities contiguous to the border in Arizona and New Mexico are also heavily dependent on underground sources. Use of groundwater in the region already exceeds normal rates of recharge.[10] Administration of groundwater use at the state level is complicated by different water laws in each of the states, and by a haphazard pattern of federal regulation. For example, state officials in Arizona and New Mexico have the power to regulate withdrawals, but in California and Texas they do not.[11] In addition, Arizona and New Mexico have adopted an overall strategy of groundwater protection, while Texas has not.

In Mexico, legal authority for water administration, including groundwater, rests with the federal government, but while sufficient formal authority to manage groundwater exists, actual enforcement is spotty. At the international level, aside from the limits of Minute 242, there are no institutionalized agreements for managing transnational groundwater aquifers.

The Mexican border economy resembles its northern neighbour in its dependence on groundwater resources. Agriculture is the main consumer of water in the region; five of the six Mexican border states are among the top ten irrigated states in the country. In the past decade, irrigated area has increased by an average of 35 per cent in these border states.

Overexploitation of several aquifers along the border will soon reach critical levels, and the groundwater problem in the El Paso/ Ciudad Juárez area is the most acute case.[12] Groundwater satisfies between 60 and 90 per cent of El Paso's municipal needs, depending on the season, and the remaining amount is drawn from the Rio Grande/Rio Bravo. Ciudad Juárez is completely dependent on groundwater for its municipal water supply. But the recharge of local aquifers is significantly less than the withdrawals. The Mesilla and Hueco Bolson aquifers together have an annual natural recharge of 29.6 million m^3, about 16 per cent of what the twin cities withdrew in 1985. Moreover, as the quantity of water stored in the Mesilla and Hueco Bolson aquifers decreases, salt concentrations increase, making the remaining groundwater progressively less usable. Another critical problem is contamination of the aquifers by pesticides and nitrates contained in irrigation run-off, leaching from waste dump sites, and dumping of untreated sewage in the El Paso/ Ciudad Juárez area.

Groundwater problems are also important along the Arizona/ Sonora border. The Santa Cruz aquifer is the major source of water for Nogales, Arizona and Nogales, Sonora. Upstream development in Mexico, and the rapidly growing water demand in Nogales, Sonora have become concerns for downstream users in Arizona. Municipal authorities in Nogales, Arizona say that water levels fluctuate widely in response to upstream uses in Mexico. The IBWC has unsuccessfully sought both surface and groundwater agreements between Mexico and the US on this drainage basin over the last decades.[13]

Water quality is also a problem in this area. The occasional discharge of raw sewage from Nogales, Sonora into the Nogales Wash, which traverses the central corridor of the two cities and flows into the US, has caused serious health problems and become a potential threat to groundwater on both sides of the border. Industrial waste from the rapidly growing border industries in Nogales, Sonora mingles with the sewage and threatens to pollute groundwater.

The two Nogales plan to continue obtaining their future water supply within the Santa Cruz Basin, but it is still not clear how long the aquifer will be able to support future economic and population growth in the two cities, particularly since no agreement has been reached on a joint management strategy. Although other pairs of border cities also share groundwater, their dependence on this source is smaller and it has not reached the same critical levels.

The growing pressure on scarce transboundary water resources along the US–Mexico border, generated by increasing water demand and pollution on both sides of the border, could create more conflicts both between the two countries, and among water users within each country. Low efficiency of use of these resources is one important element explaining these problems. Water has long been considered an unlimited resource, and little value has been placed on improving the efficiency of use. Subsidies to major water users on both sides of the border, particularly for agricultural use, have also fostered wasteful use; between 30 and 60 per cent of the water delivered to agriculture is lost due to poor irrigation practices. Significant losses are also recorded from the water delivered to urban areas (losses are 30 per cent or more in some cities). Moreover, new water demands are still met by increasing water supplies, rather than through improving the efficiency of current water use and instituting water saving programmes. Unfortunately, the disfunctionalities of the legal and institutional framework that regulates the use of transboundary water aggravates the pressures on these resources.

THE INSTITUTIONAL FRAMEWORK

The 1944 Water Treaty forms the core of the legal framework for the allocation of transboundary surface water, but the treaty does not cover groundwater due to lack of information and fears that the complexities of the groundwater question would interfere with the agreement. The treaty provides guidelines for water management and allocation during periods of 'extraordinary drought' (although this is not defined by the treaty). In the case of the Rio Grande/Rio Bravo, it stipulates that if drought prevents Mexico from delivering the required 432 million m³ average, 'any deficiencies existing at the end of the aforesaid five-year cycle shall be made up in the following five-year cycle with water from measured tributaries in which the US has the right to a 1/3 share'. In the case of the Colorado, on the other hand, the treaty guarantees Mexico 1850 million m³ per year, except in the event of extraordinary drought or serious accident in the irrigation system in the

US, at which time the water allotted to Mexico will be 'reduced in the same proportion as consumptive uses in the United States are reduced'. Note the contrast in the provisions between the case of drought in the Colorado, where Mexico is the downstream user, and that of the Rio Grande/Rio Bravo, where the US is the downstream user.

Domestic water legislation in each country is also part of the legal framework controlling transboundary water resources. In the US, intra-state allocation of water rights is basically a matter of state law. The southern boundary states use a prior appropriation system for the allocation of surface water. The basic premise of this doctrine is to give secure title to those who have undertaken to develop water resources for beneficial use by awarding water rights on a 'first in time, first in right basis'. When there are shortages, senior appropriators' rights take precedence over those of junior appropriators. Rights are generally secured through a permit process. Senior rights in the southern boundary states are held by agricultural users, while urban areas are assigned the most junior priority rights to the water. This distribution of water rights was consistent with the needs of the regional economy in the first part of this century, but the gap between this legal framework and the needs of the current regional economy has forced metropolitan areas to pursue extreme and costly measures to secure their water supplies. In addition, this appropriation system fosters inefficient use of water in agriculture, especially due to the 'use it or lose it' principle that applies to prior appropriation rights.

The legal regimes for groundwater allocation vary substantially among the southern border states. Texas has an absolute ownership common law, New Mexico a prior appropriation permit system, Arizona a groundwater withdrawal permitting process for designated Active Management Areas (AMAs), and California a combination of correlative rights and an appropriation system.

The domestic legal framework for managing resources in Mexico was established by Article 27 of the Mexican Constitution, which provides that any surface and groundwater situated in or under Mexican territory be regarded as 'national waters'. National waters may only be exploited through federal concessions, which are governed by the 1992 National Waters Act. The exploitation and use of national waters requires obtaining a 'concession' that can be in force for a period of between five and 50 years. Water rights, once they are registered, are transferable, and they can be bought and sold as long as this is technically feasible and the transfer will not adversely affect uncompensated third parties.

The institutional framework for managing water supplies is centred on the IBWC. However, despite its long tradition and its past success in managing boundary water issues, the IBWC has been unable to keep up with the growing needs of the border communities and their mounting social problems in recent years. The IBWC's approach to water management is both secretive,[14] and based on strictly technical criteria; it does not address the social disparities created by the growing imbalances in water consumption within the border communities and across the border. One of the most pressing needs is therefore the incorporation of a social dimension into the management process.

The attention the border area received during the negotiation of NAFTA facilitated the creation of the so-called NAFTA environmental institutions, including the Border Environmental Cooperation Commission (BECC), the North American Development Bank (NADBANK), and the Commission for Environmental Cooperation (CEC). However, none of these institutions has a clear mandate to improve the institutional framework for managing transboundary water resources.

The BECC's mandate is to develop and implement environmental infrastructure projects. It determines whether a project earns certification, and whether it meets certain technical, financial and environmental criteria that make it eligible for support from NADBANK, the financing institution created parallel to the BECC. BECC's first year of operation has not been easy, but it has promoted public participation and incorporated sustainable development criteria for approval of projects, both of which are encouraging signs in the agency's operation.

However, it is important to realistically assess the BECC's capacity for addressing environmental problems on the US–Mexico border, and its ability to manage transboundary water resources in a balanced and socially-grounded manner. The BECC's present mandate is really quite limited with respect to providing guidelines for sustainable development. The agency was not designed to promote sustainable development protect the environment, or manage water; it was designed as an instrument for guiding investments on infrastructure projects that impact the environment. However, BECC's mandate could evolve to incorporate a more comprehensive role for the organization in the near future if the border communities maintain their demands for further improvements in environmental protection in the area. Water issues are likely to have a central role in these demands due to the increasing gap between water demand and supply. For the local authorities and NGOs that are represented on the BECC, this might also be a useful forum for establishing a regional consensus on transboundary water issues because there is greater local input than on the IBWC.

The North American Agreement on Environmental Cooperation (NAAEC) was also created by the three NAFTA countries to promote cooperation and improve environmental conditions in North America. The governments agreed to establish the tri-national CEC, which reports to the public annually regarding the actions taken by the three parties with respect to their obligations under the environmental agreement. It may also report publicly on any matter brought to its attention, unless it is related to a party's failure to enforce environmental laws, or if the Council of Ministers affirmatively disapproves of a report. Of the three NAFTA environmental commissions, the CEC has the broadest mandate to draw attention to transboundary water issues. Although the CEC maintains a tri-national perspective in its working programme, it could provide support to improve the management of transboundary water resources at the US–Mexico border. However, it is too early to evaluate the future and the potential role of the CEC, since the CEC and the three NAFTA governments are still going through a learning process that will eventually shape the Commission's future.

Balanced growth of the border area in the future is heavily dependent on the sound use of its scarce water resources. Significant efforts have to be made to correct the inefficiencies in current patterns of water use. These efforts include upgrading the institutional and legal framework that controls the use of transboundary water resources, allowing it to respond to the new social, economic and environmental demands and needs for water on both sides of the border. This is not an easy task. The US considers reopening the 1922 Colorado River Compact and the 1944 Water Treaty to be too risky, as it is feared, for example, that strong lobbying by California would jeopardize the 1922 Compact. Agricultural users with senior water rights also fear the demands of powerful metropolitan areas. Mexico, meanwhile, fears that its position is too weak to negotiate an agreement on groundwater. However, postponing the essential effort to upgrade the legal framework – including a comprehensive review of the senior rights of agricultural water users – will only further aggravate the inefficiencies of the current system.

Institutional changes are also necessary at other levels in the water-management system. For example, water-management councils or similar organizations could be created in order to provide a forum where water users from the two sides of the border could convene, including those neglected by the present system such as environmental and Native American groups. The participation of local, state and federal authorities and other interested parties would also be beneficial for these councils. One of their major tasks would be to design sound management programmes for each basin, seeking to achieve higher efficiency in water use and conducting short and long-term planning aimed at improving equity among water users and reducing the basin's vulnerability to climate variability. In addition to local councils, a region-wide council would also be useful. This regional water council could provide a forum for the basin councils to negotiate inter-basin issues, and it could also be the place where the two federal governments could have a larger role.

A TRANSBOUNDARY WATER MARKET?

Is it possible to talk of a transboundary water market on the US–Mexico border? The answer appears to be yes, but rather than a formal water market in which Mexican water is sold across the border, what is actually evolving is an *informal* market characterized by the relocation of water-intensive activities to Mexico. While water transfers are tightly controlled in a formal market, particularly those across international borders, an informal transboundary water market allows international capital to access water resources without raising sovereignty concerns. In fact, foreign investment is encouraged and promoted by the Mexican government as a response to the new challenges of globalization. The Mexican government would face enormous opposition if it were to propose a formal transboundary water market similar to that generated by the privatization of the country's oil resources. But the development of an informal transboundary water market has drawn little attention. Rather, the high priority placed on attracting foreign capital to Mexico has left little room for questioning the real benefits and impacts of the relocation of capital.

The transboundary water market has developed for a variety of reasons, but two stand out: the increasing competition for water resources in the southwestern US due to the disfunctionalities of the legal and institutional framework and to the impact of climate variability on the availability of water; and the structural modification of the Mexican state introduced during the last two presidential administrations.

The relocation of the agricultural capital of the transboundary region from California to Baja California in Mexico provides early evidence of the effects of increasing competition for water in the US. Vulnerability to climate variability became evident in California during the six years of continuous drought suffered by the state between 1987 and 1992, which had important economic, social, political and environmental consequences. Restrictions on water use were implemented in some agricultural and urban areas. By 1991, losses in agriculture were estimated to be close to one billion dollars, as nearly 243 000 ha were left out of production, and water supply was cut by up to 75 per cent in some agricultural areas of the state. Mandatory or voluntary water cuts were implemented in the major cities, ranging from 15 to 30 per cent. At the same time, agricultural capital from California began to appear in the Mexicali Valley, and more recently, in other agricultural areas south of Ensenada, Baja California, especially in the form of *agromaquila* – that is, the rent of agricultural land (frequently communal land) to foreign capital. In 1990, up to 4000 ha in Mexicali was under *agromaquila* tenure, and by 1992, the figure was up to 20 000 ha. This rented land was mostly allocated for production of export crops.[15]

Other evidence includes several proposals to construct power-generating plants close to the US–Mexico border to export electricity to the US, and to construct a desalinization plant south of Tijuana to supply water to San Diego and Tijuana. A similar relocation of agricultural capital from Texas could be taking place during the current drought which has been affecting southern Texas and northern Mexico since 1993. Agricultural groups in both Texas and California would benefit by moving south due to lower labour and operation costs, and to the access to agricultural land with secure water rights that they would gain.

In Mexico, the structural transformation of the Mexican state, particularly during the administration of President Salinas, has been characterized by the deregulation of the federal government through the privatization of state enterprises, the reorientation of the economy toward exports, and the dominance of market-based economic principles. These transformations have included modifications to key legislation, including Article 27 of the constitution, the new National Water Act and the new National Forestry Act, which were needed to enhance Mexico's competitiveness in the global economy. NAFTA was also part of this strategy to transform Mexico. In the case of water resources, the new National Water Act allows the participation of national and foreign private capital in the management of water resources through a permit system.

For some enterprises in California, the emerging informal water market already seems to have been an effective way of increasing access to water during the recent drought period, and of avoiding the obstacles imposed by water cuts in the state. It has also been a convenient way of overriding the constraints imposed on major water users by the existing legal and institutional framework. In addition, producers may find it profitable to sell their water rights in their home state, while maintaining their production using cheaper water and labour in Mexico. However, it is important to point out that this transboundary water market is imbalanced due to the limited opportunities and disproportionate costs for Mexican users of gaining access to water resources in the US.

The intention of this chapter is not, however, to present a moral judgment on the transboundary water market or on the neoliberal policies that facilitated its creation. There is little Mexico could do at this point to reverse these policies given the constraints imposed by NAFTA, the structural transformations that the country has undergone during the last 15 years and the global dominance of neoliberalism. But Mexico could substantially reduce the negative social, economic and environmental consequences of present and future changes in the management of such key natural resource by directing more attention to these issues.

Although the formation of this informal market is in its early stages, the potential consequences for Mexico should be explored. Of particular concern is the possibility that an informal transboundary water market could lead to widening disparities in water access, not necessarily between countries but between economic sectors and social groups within Mexico. Poor water management and inequities in water consumption in Mexico already make low-income groups particularly vulnerable to these disparities. If the poor must compete with foreign capital, which has greater economic capability to lease or buy land with guaranteed water supply and to develop or obtain water distribution systems, this could lead to greater polarization in water access and water consumption. The implications for social equity, well-being and opportunities for sustainable development are not good.

Empirical evidence gathered in recent studies shows that the vulnerability to inequitable distribution of water in the northern border communities of Mexico is already higher than the official statistics suggest. For example, a study of water management and social disparities in water consumption in Nogales, Sonora indicates that the number of inhabitants that actually have access to water from the municipal system in 1991 has been overestimated.[16] Although 82 per cent of the population is hooked up to the municipal water system, water shortages are so acute that only 64 per cent of the population actually has water five or more days per week, and only 56 per cent have water ten or more hours per day. The number of people that must depend on other water sources with no quality control and that cost up to four times as much is thus at least 20 per cent higher than the official statistics suggest. Moreover, projections of future water demands for the city are being based on these overly optimistic official statistics, and they therefore carry a significant margin of error.

Nogales, Sonora shares the water resources in the Santa Cruz Aquifer with Nogales, Arizona, its US sister city. However, while Nogales, Arizona could diversify its sources of water through its access to the Central Arizona Project (CAP), Nogales, Sonora faces severe problems in maintaining its future water supply.[17] Meanwhile, average per capita water consumption in Nogales, Arizona is three to four times higher than in Nogales, Sonora, and when the disparities in distribution *within* Nogales, Sonora are considered the gap in per capita water consumption is even greater; even low-income users in Arizona use ten times as much water as the poorest residents of Nogales, Sonora.

A similar study carried out in Tijuana, Baja California in 1991 also showed that official statistics overestimated the number of people being supplied with drinking water through the municipal system. Maintenance problems were also critical; 30 per cent of the water delivered to Tijuana never reached the consumer. Meanwhile, Tijuana is the fastest growing city along the US–Mexico border, and it is already one of the largest border cities (with a population of nearly one million in 1994). It is also the city with the most severe gaps in access to clean drinking water; only 67 per cent of inhabitants had connections to the municipal drinking water system. Tijuana depends almost completely on the Colorado River for its water supply, as do the rest of the cities in Baja California. Other cities along the Mexican side of the border and in the rest of the country share similar deficiencies, but until now few studies have documented the water consumption gaps between different social groups in Mexico. A similar situation could also arise with respect to the economic use of water in agriculture, as small and medium-sized agricultural producers risk being displaced by larger domestic or foreign producers with whom they compete for water.

The deregulation of the Mexican state is already providing increasing incentives to rely on market values to manage water resources in Mexico, with potentially serious social consequences. Existing social imbalances, which are already greater than the official statistics show, may be exacerbated. These social imbalances also aggravate the competition and contradictions between the social and economic consumption of water, threatening to further depress the standard of living of many Mexicans that already live in poverty. People without access to municipal drinking water in urban areas often pay two to five times more per litre of water than those with municipal service. Residents who rely on non-municipal water supplies – which usually have no quality control – also consume significantly less water than recommended by the World Health Organization. Ironically, those with lower incomes must pay more for their water and face greater exposure to water-borne diseases than those with middle and high incomes. Moreover, due to rapid and incomplete urbanization in the Mexican border cities, even those with stable jobs and income, such as those employed in the maquiladora industry, face these problems.

CONCLUSIONS

There is no obvious short-term solution to transboundary water conflicts between Mexico and the US, but the region does seem to be developing a unique approach to long-term management of these issues in the form of an informal water market. The increasing economic interaction between Mexico and the US appears to have given business groups a leading role, thus overcoming the incentives for confrontation and conflict, and leading them instead to seek to expand and secure their access to water through relocating across the border in Mexico, and developing an informal trans-boundary water market. The deregulation of the Mexican state and its adoption of a market economy have been key factors in the creation of this market, as business people take advantage of Mexico's interest in attracting foreign capital. The potential social conse-quences of this process could be significant for Mexico, particu-larly since further polarization in water consumption would affect income distribution and the standard of living of many Mexicans. Access to water of adequate quantity and quality could increas-ingly be a deciding factor in individuals' economic and social well-being, and the gap between socioeconomic classes could widen around water issues. A similar situation could also evolve elsewhere in the world given the dominance of neoliberalism and the increase in international conflicts related to transboundary water resource issues.

This chapter also demonstrates the need to go beyond macro-level analyses that have dominated scholarly studies of transboundary resource issues. If we ignore the social, economic and environmen-tal consequences of imbalanced access to water we risk obtaining an incomplete and distorted understanding of reality, which could lead to the adoption of ineffective policies. Recognizing the im-portance of the social functions of water and the potential for con-flict with its economic functions is of particular importance. The social implications of water conflicts compel us to stop thinking of water management as strictly a technical problem, and to recog-nize that it is also a major social issue.

Notes

1. An earlier version of this article was published as: R. Sánchez, 'Water Conflicts along the U.S.– Mexico Border: Towards a Transboundary Water Market?' in *The Scarcity of Water: International, European and National Legal Aspects*, eds E.J. de Hann and E. Brans (London: Kluwer International, 1997).
2. M. Falkenmark and L. Gunnar, 'Water and Economic Development', in *Water in Crisis: A Guide to the World's Fresh Water Resources*, ed. P. Gleick (New York: Oxford University Press, 1993).
3. *Ibid.*, 99.
4. M. Reisner, *Cadillac Desert: The American West and its Disappearing Water* (London: Penguin, 1986).
5. Streamflows have ranged from a low of 6.9 billion m^3 in 1934, to a high of 30.2 billion m^3 in 1984.

6. S. Mumme, *Apportioning Groundwater Beneath the U.S.–Mexico Border, Obstacles and Alternatives*. Research Report Series no. 45, Center for U.S.–Mexican Studies, University of California-San Diego, 1988.
7. J. Calleros, 'The Impact on Mexico of the Lining of the All-American Canal', *Natural Resources Journal* 31 (Fall 1991).
8. C. MacDonald, 'Water Supply: A New Era for a Scarce Resource', in *California's Threatened Environment: Restoring the Dream*, ed., Tim Palmer (Washington, DC: Island Press, 1993).
9. D. Eaton and D. Hurlbut, *Challenges in the Binational Management of Water Resources in the Rio Grande/Rio Bravo*, U.S.–Mexican Policy Report no. 2 (Austin: Lyndon B. Johnson School of Public Affairs, University of Texas at Austin, 1992), 12.
10. Mumme, 6, note 6 above.
11. *Ibid.*, 11.
12. Eaton and Hurlbut, 41, note 9 above.
13. Mumme, note 6 above.
14. R. Sánchez, 'Public Participation and the IBWC: Challenges and Options', *Natural Resources Journal* 33 (Spring).
15. J. Bustamante, 'Subsidio Claro como el Agua', *Excelsior* (6 August 1990).
16. R. Sánchez and F. Lara, 'Estudio sobre el manejo del agua en Nogales: en busca de un enfoque integral', in *Población y Medio Ambiente: Nuevas Interrogantes a Viejos Problemas?* (Mexico: Sociedad Mexicana de Demografía, El Colegio de Mexico, The Population Council, 1993).
17. Cella Barr Associates, *Water System Study and Capital Improvement Program. Prepared for the City of Nogales, Arizona*, 1987, 19.

References

Blomquist, W. (1992) *Dividing the Waters: Governing Groundwater in Southern California* (San Francisco: ICS Press).

Gleick, P. (1993) 'Water in the 21st Century', in P. Gleick (ed.), *Water in Crisis: A Guide to the World's Fresh Water Resources* (New York: Oxford University Press).

Greenberg, A. (1993) 'The Quality of Water', in T. Palmer (ed.), *California's Threatened Environment: Restoring the Dream* (Washington, DC: Island Press).

Palmer, T. (1993) 'A Great Number of People', in T. Palmer (ed.) *California's Threatened Environment: Restoring the Dream* (Washington, DC: Island Press).

Tolba, M. (1992) *Saving our Planet: Challenges and Hopes* (London: Chapman & Hall).

Part III
Searching for Balance

14 Factors Leading to Population Growth in Bangladesh and their Impacts on the Environment

Nasim Firdaus

Correct understanding of the relationship between population growth and environmental degradation, and of the factors that lead to imbalance in developing countries, is essential to the development of an effective policy to reduce the impact of the 'population explosion' on the environment. This chapter presents a case study of the factors that contribute to population growth in Bangladesh, as well as some specific suggestions for tailoring policies to establish equilibrium.

Bangladesh is a deltaic country with an area of nearly 145 000 square kilometres (roughly the size of Wisconsin) with 230 rivers, tributaries and rivulets. Its neighbour to the north, west and east is India, while the south is delineated by the Bay of Bengal. The country also shares a small border with Burma. Bangladesh's total population has increased from 42 million in 1947, to nearly 90 million today, and some projections suggest that the population will double again by the year 2025.[1] According to a 1994 estimate, the population density is roughly 800 people per square kilometer.[2]

Despite a decline in the fertility rate since the early 1990s, Bangladesh's population growth rate still exceeds its rate of economic growth. Meeting basic needs is difficult for the majority of the population, which struggles to eke out a subsistence standard of living of two balanced meals per day and basic shelter. Under these conditions, protecting the environment remains a low priority. The lack of fuel for cooking leads to indiscriminate felling of trees for firewood, while the lack of availability of adequate food leads to overuse of fertilizers and other unsound agricultural practices.

The drive for survival has led to major migrations, both temporary and permanent, from rural areas to urban centres, creating problems of overcrowding in the cities. These overburdened urban centres are unable to meet migrants' expectations of employment and better living conditions. The results are city slums, squalid living conditions, and destruction of surrounding ecosystems as, for example, inadequate land-use planning leads to indiscriminate filling of wetlands and causes damage to wildlife and natural habitats. Meanwhile, the villages, where the majority of people still live, continue to be deprived of resources to meet their growing needs.

What are the causes of this rapid population growth which puts such tremendous stress on the country's resources and overwhelms development efforts? The answer lies in a number of interlinked economic, social, cultural and psychological factors.

FACTORS DETERMINING FAMILY SIZE

Eighty per cent of Bangladesh's population lives in rural areas where the primary occupation is agriculture. Although the contribution of agriculture to the overall GDP has declined significantly over the past few years, this sector still employs more people than any other. Of the total land mass of 13 million hectares, 9 million is arable.[3] For the rural poor, maximizing profit from agricultural land is necessitated by the low level of industrialization and the absence of a diversified economic base and employment alternatives.

In land-scarce Bangladesh, manpower is the only input available to intensify and extend cultivation. Because agriculture is so labour-intensive, women are obliged to have many children. Additional manpower also allows poor families to diversify their sources of income. The negative consequences of large families, however, include fragmentation of landholdings and reduced farm size, as well as increasing landlessness.[4] These consequences feed into a downward economic spiral that leads to abject poverty. As more and more farmers are marginalized, the limited economic and commercial base further restricts income-generation opportunities, resulting in spreading and intensifying poverty. Poverty in the rural and urban areas prevents families from educating their children, which in turn contributes to joblessness, hunger and unabated reproduction.

In addition to economic variables, a number of social, cultural and psychological factors play important roles in Bangladeshis' decisions about family size. Most importantly, males are seen as the income-generating gender. In addition to working in the fields, they can migrate to more affluent areas to seek work and contribute to the survival and security of their families. Male children thus provide social and economic security for their extended families through the interlinking of generations. It is the primary responsibility of the oldest male child to look after his parents, grandparents, unmarried brothers and sisters, and sometimes other older family members. They provide social and economic security for the older generations in return for the investments their elders made in them when they were young. As a result of the fact that there is so little money to live on in the present, the concept of saving for the future is almost non-existent among rural middle-income and poor families. If and when there is 'extra income', it is spent on educating male children.

The status of women, on the other hand, is very low because they are seen as burdens. Without power or prestige, and with little autonomy, they rely on men – husbands, brothers or sons – for their well-being. Due to their dependent position, women seek to have many sons, because they will both protect them against threats of divorce and desertion and become their economic, moral and social-psychological support in the future. Because of the social, economic, and cultural preference for male children, women who fail to give birth to male offspring are often threatened or made victims of divorce, discretion or a husband's remarriage. As a reaction to this, women tend to acknowledge the desirability of having male children – often more than one male child to ensure the survival of male descendants – and they are often expected to produce children throughout their fertile years.

Traditionally, men have believed that women are subordinate and inferior creatures, a belief that is also accepted by most women. Girls are deprived of equal opportunities and other social benefits such as formal education and employment. Males in the family eat first and receive the best of what is available, especially during periods of scarcity, while female members are often unfed or underfed. Deprivation for women often begins in childhood, and many remain malnourished throughout their lives. Daughters are only expected to care for the emotional or physical well-being of their parents and the other dependents of their brothers. It is otherwise socially unacceptable for parents to become dependent on their daughters in any other way. Parents will not live in their daughters' homes unless they are compelled to do so because they have been unable to produce a male child.

While the purpose of having a male child is to ensure economic and social security, most poor couples having given birth to many children have often unconsciously added to their economic burden. Because of the precariousness of life in Bangladesh, couples also tend to have many children to ensure more surviving children, to allay their own fears and to meet the insistence of elders within the society. Even with improved health care and immunizations, the death rate for infants and children remains high, in part because undernourished or malnourished mothers have babies with low birth weights. The limited access to health-care facilities also means that more children suffer from continuous illness.

Having many children is also valued because of the cultural belief that growing up with siblings helps children to acquire essential social values that cannot be acquired outside of the family. Sharing feelings, emotions and responsibilities among siblings builds the foundation of emotional stability in the society. Larger families are thus seen as assets rather than liabilities, at least by many parents.

Religious law also guides most personal life, including inheritance laws which favour male children. In the case of a husband's death, the widow inherits one-eighth of the husband's land if they have had a child. However, if there is no child the deceased husband's land is turned over to his family, and his widow is left with nothing (a husband gets exactly twice as much in the case of his wife's death). If a family has had only daughters, the children are deprived of a large share of their father's property because it will go to the deceased husband's brothers or other male family members. In all cases, sons receive double what daughters receive. Since sons are so favoured, it is natural for women to prefer sons; as inheritors of their father's property they are expected to provide security for their mother and sisters.

Dowries also have an influence on family decisions. To avoid the burden of providing a dowry and the possibility of violence against daughters if the dowry is insufficient, parents often prefer to have sons. Male children can also assist in providing a dowry for their sisters. Thus, if a couple has several daughters they are likely to continue trying to conceive a male child, regardless of how large their families become.

MAKING FAMILY-PLANNING AND POPULATION MANAGEMENT EFFECTIVE

The international community has placed heavy emphasis on the use of contraceptives and family-planning in its efforts to stem the tide of population growth in the Third World. Although Bangladesh has been more receptive to these measures than many of its Muslim neighbours, it should be pointed out that this emphasis represents an oversimplification of the problem. One study has shown, for example, that despite widespread availability of contraceptives in Bangladesh, they are not widely used.[5]

Significantly, there will be millions of Bangladeshi women of child-bearing age in the coming decades who will not have the power or authority to make their own decisions concerning family size. It is therefore essential to focus on women as individuals and on the economic, social and cultural factors that have encouraged women to have large families; that is, the need for labour in agriculture, social security for women, gender bias and an overall lack of development.

Elimination of poverty is a long-term project, yet this is an essential element for the welfare of global society. The immediate short-term goal is the alleviation of economic distress through investment in industry and through development projects that can stimulate economic growth. Agriculture will continue to remain the major sector of employment since it has the capacity to absorb excess labour more easily than any other sector. Strengthening the agricultural sector requires not only improved technology for higher and better yields, but also widespread development of infrastructure such as roads, storage facilities and transportation systems. If the rural areas are linked, migration will slow down, stimulating a decline in population growth.

Removal of trade barriers on exports from Bangladesh, especially the removal of import-quota restrictions imposed on ready-made garments by the US and other Western countries, would also alleviate the situation. The majority of the workforce in the garment industry, the fastest growing non-traditional export base in Bangladesh, is composed of poor rural and urban women. Imposition of import quotas has forced closures and production slow-downs in the industry, which has the capacity to provide women with the economic and social security essential for enhancing their power and their role in determining family size.

Enhancing women's status encompasses more than providing equal access to education, health care and employment. It also requires increasing their prestige, power and autonomy. There is no doubt, however, that investment in education, health and even agriculture can contributes to the enhancement of women's power as it helps the younger generation of women to gain entry into schools and find employment. There are, however, millions of women of child-bearing age who cannot benefit from these policies. It is essential to provide these women with a sense of security if there is to be a decline in the fertility rate. Enforcement of the existing family laws that are favourable to women can also act as an effective deterrent to divorce and desertion on the grounds of children's gender.

The introduction of health care and social security provisions for certain groups of women, for example those from 25 to 60 year's old, would begin to erase the cultural and social 'stigma' attached to being female. The threat of divorce and polygamy will diminish as men begin to realize that women have 'power', and that they can be an asset rather than a burden. Social security could be provided through a monthly 'honorarium scheme' that would provide recognition for women's contributions to society's well-being. Today, mothers and would-be-mothers require maximum attention, yet they are now generally deprived of health services.

If the proposed honorarium is limited to women from 20 to 60 years of age, then investment for fewer than 20 million women, or less than 28 per cent of the total population, will be required. Providing a relatively small amount of funds to women is likely to result in micro-level investments and income generation. As the experience of the Grameen Bank (which provides credit to rural women for microenterprise) has shown, the entrepreneurial efforts of rural women can be extremely effective and dependable. The primary objective of the honorarium scheme should not, however, be microenterprise financing, but socially-oriented equity and economic empowerment. Any kind of economic empowerment, no matter how small, will have substantial impact on the society and will very likely raise the prestige and status of women. This programme can be introduced in phases on the basis of age, economic level or employment, or a combination of all or some of these factors.

In summary, conscious efforts to raise the status of women would give women more freedom and help them choose their life options, including the number of children they will raise. These efforts would, however, have to be part of an integrated approach that encompasses other economic and social interests.

Global fears about overpopulation reflect concerns that humanity's exploitation of the earth's environment is becoming unsustainable, and that this will result in an increasing threat of international migrations. Where there is international pressure on developing countries for a quick reduction in population growth, it is important to recognize the various factors that lead to population growth in developing countries like Bangladesh, and the necessity of allowing these countries to develop to their full potential. Populations that can support themselves in their local environments will have no desire to migrate to a foreign land.

Notes

1. Government of Bangladesh, *Statistical Handbook of Bangladesh* (Bangladesh: Government of Bangladesh, 1991).
2. Planning Commission, Ministry of Planning, Government of Bangladesh, *Planning Draft for 1995–2010 for Nipharmi District*, July 1995.
3. Government of Bangladesh, *Statistical Handbook.*
4. R.H. Chaudhury, 'Population Pressure and its Effects on Changes in Agrarian Structure and Productivity in Rural Bangladesh', in *Population Growth and Poverty in Rural South Asia,* ed. G. Rodgers (New Delhi India: Sage Publications, 1989). In 1977, the average farm size in Bangladesh was 1.4 hectares (ha), but by 1983–84 it had dropped to only about 0.9 ha. Moreover, according to some estimates, while 35 per cent of rural households owned less than 0.2 ha of land in 1960, by 1992 this figure had risen to 53 per cent. As a result, many small farmers are unable to produce enough crops to sustain their families.
5. K.V. Rangan, 'Population Services International Abridges the Social Marketing Project in Bangladesh', Kasturi V. Rangan, 'Population Services International (Abridged): The Social Marketing Project in Bangladesh,' Harvard Business School Case Study, Publishing Division, Harvard Business School, Boston, MA, 1990.

15 Two Threats to Global Security[1]
Alex de Sherbinin

In October 1347, a Genoese merchant ship returned from a trading expedition to the Black Sea, docking at Messina in Sicily; on board the ship were sailors who had been infected by the bubonic plague. The plague had originated in China, which at the time was one of the world's busiest trading nations. The epidemic followed a path along the silk road to the Black Sea, Europe's gateway to the treasures of the Orient. From there, the Italian merchants brought the deadly cargo to Sicily and southern Italy. By the following year – approximately 15 years after its initial outbreak in China – the plague had spread as far north as England, where it was dubbed 'the Black Death' for the black spots it produced on the skin. Within five years the plague had claimed one-third of Europe's population (25 million people), significantly altering the course of European social and economic history.

Obviously, this factual account of the plague's spread does not do justice to the huge toll it took in terms of human suffering, and the tremendous fears it generated in European towns and villages as their inhabitants wondered when and where the plague would strike next. Lacking an understanding of the principal host and the vector, rats and fleas respectively, medieval Europeans could do little to control the epidemic. Environmental sanitation conditions were appalling, and would not see significant improvement until public health measures were first implemented in England four to five hundred years later.

The plague constituted the western world's first environmental security calamity. In relative terms, if the same epidemic were to occur today it would eliminate the equivalent of the entire US population. Just as the plague was a bacterial infection that took advantage of patterns of trade and alteration of the environment to spread itself, today's environmental security threats are likely to come from nature's responses to human-induced changes in the planet's biophysical systems. Although today we have a knowledge of the workings of the natural world undreamt of in the Middle

Ages, the environmental challenges confronting us are also several orders of magnitude larger. The world is becoming warmer due to greenhouse gases, more fertilized due to nitrogen fixing, and drier due to human appropriations of fresh water and changes in land cover. Furthermore, the number of people living in miserable unsanitary conditions that can only be compared to those of medieval European cities is also far greater than at any time in the past.

Population growth and patterns of economic globalization (as manifested by consumption, trade and the growing power of multi-national corporations) may have grave and unforeseen consequences for the environment and humankind. As with the plague 650 years ago, there is a sense in which these processes are seemingly beyond our control, as if the root causes of the current global disequilibrium are not sufficiently understood to allow us to reverse the trend. Although a growing number of people have a vision of an alternative future based on simpler lifestyles, greater equity, meaningful work and social justice, getting from 'here' to 'there' will be tougher than most imagine. Still, it is necessary for humanity to choose its future, or the world will continue to be beholden to processes that perpetuate vast inequalities and generate tremendous human suffering.

THE STARTING POINT: POPULATION, AFFLUENCE AND CONSUMPTION

> The poor are locked in poverty largely because the rich control the world's markets, resource flows, prices, and finance. But they are aware of one another. Modern communications and tourism bring the luxury of the rich before the eyes of the poor, and the latter no longer accept these disparities with patience or as a part of some natural historical order.[2]

The greatest threats to global environmental security stem from the activities of two particular subsets of the global population: the absolute poor and the affluent. These sub-populations cut across national borders, and do not strictly correspond to the traditional regional categories of lower, middle and upper-income countries. Thus, a poor indigenous person in Malaysia or Brazil, both rapidly industrializing countries, would be on the same socioeconomic level and have essentially the same environmental impacts as a rural farmer in Mali, one of the world's least developed countries. Likewise,

Table 15.1   Comparative estimates of world population
by wealth category

UNDP – 1993[3]	Billions	A. Durning – 1992[3]	Billions	IIASA – 1990[4]	Billions
Richest fifth (83% of income)	1.1	Consumers	1.1	Top	0.29
Upper-middle (12% of income)	1.1		3.3	Upper-middle	1.00
Middle (2.3% of income)	1.1	Middle inc.		Lower-middle	
Lower-middle (83% of income)	1.1			Bottom	
Poorest fifth (1.4% of income)	1.1	Poor	1.1		
Totals	5.5		5.5		5.29

Source: UNDP, *Human Development Report 1992* (New York: Oxford University Press, 1992), p. 35. A. Durning, *How Much is Enough?* (New York: W. Norton, 1992), p. 27. W. Lutz, C. Prinz and J. Langgassner, 'World Population Projections and Possible Ecological Feedbacks', *Popnet: Population Network Newsletter* 23 (Summer 1993), table 4, p. 10.

with some minor differences, the up-and-coming computer pro-grammer in Bangalore, India, and the middle-income residents of Madrid, Spain are on a par with respect to their consumption of resources and waste generation. Although precise numbers for the populations in these two sub-groups are not generally available, a reasonable estimate is that approximately 1.5 billion of the world's population are very poor, and another 1.5 to 2 billion are affluent (for example, the UNDP estimates that 95 per cent of the world's income is controlled by the wealthiest 2.2 billion members of its population; see Table 15.1 for some other estimates). With today's world population of 5.9 billion, this leaves about 2.8 billion people in the middle-income category (Population Reference Bureau, *1998 World Population Data Sheet*, Washington, DC).

The absolute poor generally act in environmentally threatening ways at a very local level, through practices such as over-cultivat-ing, over-harvesting and deforesting land, encroaching on protected areas, clearing land on steep slopes and disposing of wastes in an unsanitary manner. They do so out of necessity, and because they often lack the capital, appropriate technologies and training to adopt alternative practices. The threat they pose stems partly from their practices, and partly from the scale and magnitude of these prac-tices which are directly related to the increasing numbers of poor people. This is a case of local-level environmental degradation occurring on a large enough scale that, in aggregate terms, amounts to a global problem (at least in the developing world).

In the final analysis, however, the material 'needs', consumer preferences and economic activities of the affluent pose a far greater threat at the global level. These needs, preferences and activities are directly related to culturally prescribed patterns of consumption and the ongoing process of economic globalization. Free trade, the dominance of multinational corporations, and the models of development that are currently being pursued are creating a situation that will have severe long-term environmental implications. Again, it is a question of scale. Although the affluent are a relatively small subset of the world's population, they consume a disproportionate share of the world's resources, and their numbers are growing rapidly as countries around the world strive to compete in the global economic marketplace for their share of the economic pie. This chapter will therefore focus on the environmental impacts of the affluent.

Much of the developed world's attention and policy action has been focused on the destructive impact of the world's poor, and on the security threats that unbridled immigration from poor countries poses to the US and Europe.[3] Action to address environmental problems and rapid population growth amongst the absolute poor in the developing world is indeed necessary, and with reasonable resources many of these problems can be alleviated. However, actions taken in this realm have often served to divert attention from environmental problems related to current trends in consumption, trade and development that altogether constitute 'economic globalization'. These problems include global warming, ozone depletion, over-fishing, deforestation and hazardous waste disposal. Both industrialized and developing countries share responsibility for these problems because, as stated above, the affluent live in both the North and the South. But it is the industrialized world through its disproportionate consumption of natural resources, production of greenhouse gas emissions and hazardous wastes, and dominance of world trade, that is having the greatest environmental impact.[4] As will be discussed in the sections below, these problems are far more intractable because they involve complex social and institutional arrangements and widely-shared assumptions regarding lifestyles and the benefits of economic growth. In short, addressing these problems requires an examination of the very assumptions upon which the current world system is based.

THE CONSUMPTION–ENVIRONMENT NEXUS

Colourful depictions of 'ugly' over-consumers – for example wealthy Americans who drive ostentatious luxury cars, fly in private planes, own yachts, wear animal furs and indulge their tastes for exotic foods imported from around the world – have tended to cloud understanding of the consumption–environment nexus. Many people believe that if society could simply eliminate such extravagant and wasteful consumption patterns, and make an extra effort to drive less and recycle cans and bottles, that many environmental problems would be resolved. This betrays a basic misunderstanding of the size and scope of current environmental problems, and of how difficult it will be to change the consumption and behaviour patterns of average consumers in the industrialized and the newly-industrialized countries in order to have a meaningful impact on the environment.

Between 1960 and 1995, global population nearly doubled from 3 to 5.7 billion, and global economic output more than tripled from $5.7 trillion to over $19 trillion (in 1987 US dollars). Between 1995 and 2030, world population is likely to grow by half again, to 8.7 billion, and global economic output is projected to increase three-and-a-half times, to approximately $67 trillion.[5] Rapid growth in the economies of East Asia, Latin America and the Middle East, coupled with a possible tripling of the economic output in industrialized nations, accounts for most of this projected growth. The nascent consumer class in the countries of south and east Asia already outnumbers that of western Europe and North America by several million. Citizens in these Asian countries, and in most other developing countries, apparently wish to emulate the living standards of the developed world, and are being encouraged to do so by media advertising campaigns.[6] Although the environmental repercussions of this rapid economic growth will depend on the nature and composition of global economic activity and the technologies employed, the projected increases in the production and consumption of goods will undoubtedly have a significant impact. In other words, the energy use and material throughout of the economy will not grow in direct proportion to gross domestic product (GDP), but it would take major advances in science and technology to decouple them entirely.

At a superficial level, consumer demand is what drives the production of goods and services, which in turn generates environmental impacts. Indeed, under the assumption of consumer sovereignty, economists identify the market as the specific mechanism by which goods and services are produced in response to consumer preferences.[7] This might lead us to conclude that individuals and households have the greatest responsibility for the environmental effects of consumption, which is only partially true. While there is no question that 'we' as individuals are solely responsible for our consumption-related decisions, this simple market-based model does not take into account a number of important factors that make the relationship between consumption and the environment more complex. The first, mentioned above, is advertising. Over the course of this century, marketing specialists have developed highly sophisticated means of advertising to shape consumer preferences and stimulate demand for products that meet psychological needs (as opposed to basic needs for shelter, food, clothing and transportation). This is a case of the market 'tail' wagging the consumer 'dog'; if it were not for large-scale advertising campaigns, many of the products that are harmful to the environment (and to human health, such as cigarettes) would not have such widespread appeal.

The second factor contributing to complexity relates to government procurement and purchasing patterns. There are a whole set of market demands that are generated by governments and large institutions, for products ranging from large buildings and power plants, to military hardware. Moreover, governments also have the power to intervene in markets through regulatory action and the provision of subsidies, and they thus bear a significant responsibility for the impacts of consumption on the environment.

The last of the complicating factors relates to the design and production of consumer products. Decisions made within corporations about which industrial processes or materials to use for the production and packaging of certain goods may only be marginally influenced by consumer preferences.[8] These choices, for the most part, are made by engineers, scientists and research and development specialists, in conjunction with marketing teams, within limits imposed by government regulations. Provided that the end products are essentially the same, consumers may be indifferent to these 'embedded choices' that form a part of the product design, production and marketing processes, even though they may have significant environmental implications. For example, the styrofoam packaging used by McDonald's was eventually replaced by paper packaging when environmentalists pointed out that styrofoam contributes to CFC emissions and is non-biodegradable. From the perspective of the consumer, the current packaging works just as well – the company's marketing and research and development staff simply had not initially considered the environmental implications of their packaging choices. The point here is that these factors have little to do with the consumer preferences of individuals and households, and yet they add tremendously to the complexity of the consumption–environment nexus.

At the dawn of the industrial revolution two centuries ago, when Adam Smith wrote *The Wealth of Nations*, markets were primarily local. The person who produced your horse-drawn cart was local, and would probably repair the cart for you as well. This is no longer the case. Instead, goods, and increasingly services, are often produced at great distances from where they are consumed. Tomatoes and lettuce produced in California are shipped by truck and air freight around the world. Even the water consumed by élites in Singapore, Hong Kong and throughout the US is bottled in France. Services are also increasingly performed at a distance. Insurance claims generated in the US are sent by fax to Ireland to be processed. Capital is free to go wherever labour is cheapest and environmental regulations are most lax. This is the essence of economic globalization. The global economy knows few boundaries, and the corporations and interests that benefit from economic expansion are rapidly seeking to knock down the few barriers to trade and to the movement of capital that remain.[9]

DRIVING TO DISTRACTION

The process of economic globalization and its links to consumption are well illustrated in the case of automobiles. The automobile is a supremely useful invention, a fact that is confirmed by the consumer choices being made around the world. More than a means of getting from point A to point B, it reflects broadly held values such as the importance of individual freedom and mobility, and provides a vehicle (literally) for personal statements of prestige, power, and even environmental awareness (witness the small economical models). In Jungian dream analysis, cars are symbolic of personal power, ego and drive. However, humanity's love affair with the car is having major environmental ramifications. These take the form of greenhouse gas emissions (CO_2 and CFCs), urban smog,

Figure 15.1   World passenger car registrations and production, 1950–94

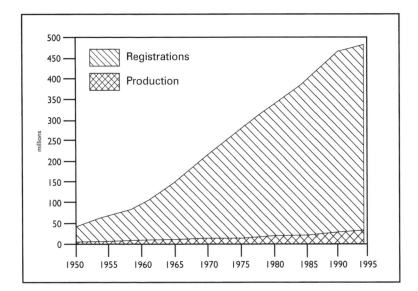

Source: American Automobile Manufacturers Association (AAMA), *World Motor Vehicle Data 1996* (Washington, DC, 1990) pp. 2–15.

noise pollution and congestion, the paving of once productive or aesthetically pleasing landscapes, and depletion of petroleum reserves. There are additional environmental costs related to the manufacture (mining of raw materials, energy use and chemical disposal), maintenance (for example crank case oil, CFCs for air conditioners) and disposal of automobiles and their component parts (batteries, tires, air conditioning units).

The scale of automobile use is increasing at an alarming rate around the world, and this is occurring well beyond the traditional strongholds of North America and Europe. Between 1950 and 1995, the total number of passenger car registrations increased at an average annual rate of 5 per cent, from 53 to over 500 million (see Figure 15.1). In 1950, three-quarters of all registrations were in the US, while in 1995 this proportion had declined to 30 per cent. The production of passenger cars has remained steady at an average rule of 35 million cars per year for the past decade, up from 8 million in 1950. The population per car ranges from two in North America, four in Europe and 15 in South America, to 69 in Africa.[10]

Globally, one-third of world oil consumption and 14 per cent of carbon dioxide emissions can be attributed to motor vehicles. In the US, the figures are 50 per cent of oil demand and 25 per cent of carbon dioxide emissions.[11]

On average, Americans consume roughly 22 litres of fossil fuels per person per day, generating 19.5 tons of carbon dioxide annually per person.[12] If the rate of fossil fuel consumption were to be reduced to sustainable levels (that is, no net contribution to the greenhouse effect), average consumption per person worldwide would need to be reduced to 1 litre of carbon-based fuel per day by the year 2010. Excluding all other fuel use, an individual would have the choice of traveling only 24 kilometers (km) by car, 50 km by bus or 10 km by plane each day.[13] We are participating in a massive global experiment in which each year billions of gallons of fossilized carbon that took millions of years to form are being extracted and pumped into the earth's atmosphere. No one can accurately predict the full environmental consequences, though climate change, increasing seasonal and spatial variability in rainfall, sea level rise and displacement of ecosystems are among the results that scientists reasonably expect. This may cause a return to the kind of jungle-like conditions that produced the fossil carbon in the first place.

It appears that in this domain, as in many others, the imperatives of the financial world are taking precedence over the needs of the living world of humans and other species.[14] Car manufacturing and sales are big business. With annual sales of around one trillion dollars, the industry accounts for one in ten jobs in industrialized countries. According to *The Economist*,

> Developing countries, having seen the wealth the industry has created, aspire to their own car plants as symbols of economic virility and a source of jobs. With markets in America, Japan and Europe now growing only slowly, the world's big car companies are scrambling to set up shop, with local partners, in China, India and Latin America, where car output this year [1996] will rise by a quarter.[15]

In other words, large-scale marketing campaigns and agreements between developing-country governments and car manufacturers are encouraging the spread of US-style transportation models all over the world. For example, the *New Scientist* reports that in Calcutta, the Japanese International Cooperation Agency is financing the construction of six highway overpasses to the tune of $46 million, which is more investment than the city's tram system has received in the past 20 years.[16] Bicycle rickshaws and hand-pulled carts will be banned from the centre of the city, and tram-lines that cross busy intersections will be shut down, all in favour of a mode of transportation that will solely benefit the middle and upper classes, while increasing air pollution and creating havoc in the streets.

Scenarios like these are becoming commonplace in the developing world. Bangkok is so congested that traffic crawls along at less than 10 km per hour. Even in Africa, where car ownership is most limited, cities such as Lagos, Nairobi and Dakar have fallen prey to the craze for personal transport, paralyzing their streets with grid-lock. In China, where economic growth rates have averaged close to 8 per cent per year over the past decade, the growing middle class is hungry for automobiles and other consumer goods. The number of automobiles is expected to increase 11-fold between now and 2010, from 1.8 million to 20 million.[17] With the ubiquitous bicycle being replaced by cars, the country is headed for an ecological catastrophe that will have global ramifications, not to mention monumental traffic jams.

In a world in which the consumer is sovereign, the collective impact of millions of individual decisions in favour of personal transportation is leading us incrementally away from environmental sustainability.[18] As incomes rise, people desire more individual freedom and mobility. However, if everybody paid the true environmental and social costs of personal transportation using the automobile (that is, if these costs were 'internalized' via higher sticker prices, gasoline taxes and tolls for road use), then cars would be a far less attractive option when compared with public transportation. For those who could still afford them, cars would be more fuel-efficient and environmentally friendly. But that day is still far off. The successful battle against the BtU (British Thermal Unit) tax in the US shows how vigorously the automobile and petroleum industry trade associations fight against any proposals that would increase the costs of personal automobile use. Today, most environmental and social costs of the personal automobile are externalized (or socialized, via government subsidies); furthermore, even if they were to be internalized, the true long-term costs of large-scale transportation based on the personal automobile are not fully known. Unfortunately, the prospects for moving the world towards a more sustainable transportation model are growing dimmer with each additional kilometer of black top that is laid.

SURVIVAL OF THE FITTEST

The forces that pursue the expansion of global economic activity, whatever the perils to the environment, are far stronger than the countervailing environmental forces. Societies around the world are increasingly being shaped by a form of economic Darwinism in which only the fit survive. Global competition in the name of greater economic efficiency is the holy grail of economic globalization. This raises important questions, however, about who truly benefits from this globalization process. Cornucopians and other proponents of unbridled economic growth would have us believe that everyone benefits from greater production of material goods, because more goods implies greater well-being.[19] However, others are beginning to question this basic assumption of the modern world system.

Sociologists W. Kempton and C. Payne address what they term the myth of 'social evolution for individual benefit'. By myth, they are referring to a set of ideas that persists and spreads because it has some function, or fits within culturally prescribed belief systems, and thus is commonly accepted. The 'social evolution for individual benefit' myth is based on the idea that the organization of human society is going through an evolutionary process that increases the benefits (materially and psychologically) to all of its members. However, upon reviewing the evidence on the two major revolutions in human organization (that is, from hunter-gathering societies to sedentary agricultural societies, and from agricultural societies to industrial ones), they conclude that 'major social transitions will occur if they provide benefits to decision-making elites and greater "fitness" at a societal level (that is, military advantage or rapid growth and spread of the socio-political system). Increasing the quality of life of individuals is not a criterion for choosing the direction of social evolution'.[20]

In the current social transition, defined as it is by increasing globalization and the growth of multinational corporations, this principle is borne out once again. The concentration of wealth and power in the uppermost echelons of society continues unabated, while a large and growing number of people are losing out.[21] Even average Americans, who are the world's winners from the perspective of per capita income, find themselves on an economic treadmill that cheats them of some of the basic components of a quality life: free time, low stress, low crime and a happy family life.[22] Several million Americans now find themselves among the working poor, frustrated by jobs that pay just above the minimum wage, with few benefits and even fewer prospects for advancement.[23]

In the developing world, a small number of people are benefiting from economic globalization, largely at the expense of the poor and economically disenfranchised. In the words of M. Castells,

> [t]he current process of restructuring is fragmenting the social fabric of the planet. It is also recomposing it, but only in part, into a structure that primarily suits the interests of dominant governments and corporations, and of those areas or institutions for which the dominant powers have specific interests.[24]

Proponents of free trade as institutionalized under the World Trade Organization argue that developing countries can benefit from trade liberalization by taking advantage of their comparative advantages. However, the principal comparative advantages of most developing countries – plentiful supplies of cheap labour and lax environmental regulations and enforcement – are more likely to lead to social, economic and environmental impoverishment than to improved well-being.[25] In fact, many poor farmers and labourers in the developing world have already experienced first hand the 'benefits' of free trade, as they see their agricultural lands expropriated for commercial agriculture, or they find themselves in assembly-industry jobs earning survival wages while producing goods for Northern consumers.[26] In short, the quest for greater efficiency in response to global competition has not produced greater human well-being, but has instead led to progressive dehumanization.

A RESPONSE TO THE CORNUCOPIANS: WHY TECHNOLOGY AND INFORMATION WON'T 'SAVE' US

When discussing global environmental trends, cornucopians and other proponents of economic growth have tended to focus on the proposition that unlike other animal species, humans have culture and knowledge and can therefore shape their environment to a degree that far surpasses any other living organism. As Lutz notes, the very fact that humans think about their own impact upon the environment sets us apart from other species.[27] The fundamental question regarding the fate of the planet can thus be reduced to the following: will human culture be able to generate responses to the very environmental problems that humans are creating fast enough to remain 'in control' of the process, or will there be a point at which humanity has gone too far, and finds that environmental (and even societal) processes are beyond its control?

The standard response of the cornucopians and those who benefit from current world economic arrangements is that science and technology will develop solutions to many of the problems of diminishing resources, and that the information explosion will accelerate the development of these new technological approaches. It is indeed true that science and technology have served society well. Today, thanks to the green revolution in agriculture more people are being fed than ever before, and due to medical advances there is less sickness and disease and life expectancy has increased dramatically. Via the Internet and international courier services, people are now able to communicate faster than at any other time in history.

However, these dramatic claims regarding the future impact of science and technology have a disconcerting element of idolatry, for they are based on the supposition that man can save himself from his increasingly perilous situation if he just applies sufficient know-how. This is science and technology as religion. Ironically, it is not hard scientists who evince the most faith in science and technology; rather, it is usually cornucopian economists. A recent joint statement from the scientific academies of 58 countries makes this point clearly:

> As scientists cognizant of the history of scientific progress and aware of the potential of science for contributing to human welfare, it is our collective judgment that ... it is not prudent to rely on science and technology alone to solve problems created by rapid population growth, wasteful resource consumption, and poverty.[28]

Rustum Roy, a mechanical physicist, writes that '[h]umanity's capabilities are bumping up against the absolute ceiling of Nature's givenness'.[29] Despite abundant evidence that progress in science and technology is reaching a plateau as it confronts the limits imposed by the fundamental physical principles governing the universe, Roy laments that most US government committees on research and development policy still plan within the framework of a linear growth and progression hypothesis. In the field of biology, where new revolutions are still occurring in the area of genetic engineering, it is likely that scientists will run foul of widely held ethical values regarding the sanctity of life, resulting in a slow down – if not a halt – in further progress.

In contrast to these modest statements by scientists, Cato Institute economist Stephen Moore writes,

> As we progress further into the current information age, the notion of finite physical resources is becoming all the more outmoded. At an unparalleled pace, human ingenuity is unlocking ever more spectacular advances in technology and scientific knowledge that are advancing our mastery over the finite physical universe.[30]

In essence, the cornucopians project a rapid growth in scientific discoveries on the basis of the information revolution and past trends in technological progress.

The information age has indeed brought a deluge of information, some of which has been quite useful for science. Aside from television (which is the most popular medium of information dissemination but contributes little to scientific advance), there has been an incredible upsurge in information exchange made possible by advances in telecommunications. For example, Internet connections have increased from a few thousand e-mail accounts in 1980, to between 30 and 40 million today, and the number of sites on the World Wide Web is up from 130 in 1993, to 650 000 in 1997.[31] Within a few years it will be possible to find almost any information on the Web (given sufficient time to sift through what is not useful). There is also no question that communication and the exchange of ideas between scientists has been enhanced by electronic mail.

The catch is that an increase in information does not mean a commensurate increase in our collective well-being. As we become inundated by information and facts from our newspapers, television news and Internet sources, we need to fit them into some kind of framework in order to make sense of them so that they will contribute to the development of our world view and to the advance of civilization. As E.F. Shumacher wrote,

> We cannot get an overall view merely by assembling more and more facts. By themselves, facts mean nothing, prove nothing, and lead to no conclusions. Facts need to be evaluated, that is to say fitted into a value system, to be of use.[32]

More and more information generated at greater and greater speed, without time for reflection or putting it in its proper context, simply generates information overload. This is knowledge as junk, an accumulation of dissociated 'factoids'. Though we try to order the information better through information management and databases, the sheer scale of the task is quite daunting. Fundamentally, unless we change the value system and assumptions upon which the global economic system is operating – pursuing economic advance for the sake of human welfare, instead of for personal enrichment, material gain or to further the aims of corporations – we will just harness increased information and technological progress for the same ends.

TACKLING THE TWO THREATS: CREATING ALTERNATE FUTURES

For now and the foreseeable future, the greatest threat to the environment stems from 'economic globalization', which is directly related to trends in consumption and trade, and to the currently available models of development. Attempts to put the world on a more socially and environmentally sustainable course through population stabilization and environmental protection are necessary and need to be pursued more vigorously. The problem is that the sums of money invested in pursuit of these goals are dwarfed by the size of private-sector investments that, each day, are incrementally moving the world away from environmental sustainability. Furthermore, it can be argued that the aid-dollars dedicated to population stabilization and environmental conservation serve as red herrings, diverting attention from the far more intractable issues surrounding economic globalization.[33] Although efforts in these two domains *do* need to be stepped up, commitment to these activities on the part of governments, multinational corporations, international non-governmental organizations and the UN cannot be used as an excuse for inaction on other more vital fronts.

Paul Harrison, a British writer on population and environmental sustainability, writes of a 'Third Revolution'.[34] The first revolution was the shift from hunting and gathering to agriculture, with all the attendant changes in settlement patterns. This permitted the specialization of labour, the rise of large civilizations, increased concentrations of population and a whole host of other societal changes. The second revolution was the industrial revolution, with its incredible reliance on petrochemicals. This has resulted in further specialization of labour, even more complex social organization, and the rise of megacities of 10 million people and more.

It is still unclear what the third revolution is going to be. There have been attempts during the recent series of UN conferences (UNCED, the Human Rights Conference, the International Conference on Population and Development, the Social Summit, the Women's Conference, and Habitat) to create action plans to redress global inequalities, stabilize population and improve the environment. They hint at a more equitable, just and sustainable global society, but they suffer from two weaknesses: they lack teeth, and they fail to address the underlying causes of inequity and unsustainability which can be found in the process of economic globalization described above.[35] Moreover, being gatherings of government officials representing nation-states, they may also lack the imagination or foresight to recognize the next major revolution. Perhaps this is because the nation-state itself may become a thing of the past.

Although we do not know what society will look like after the third revolution, it can only be hoped that it will demonstrate some of the following characteristics:

- a respect for human beings, especially disadvantaged groups such as the poor, children, the elderly, ethnic minorities and women;
- a willingness to challenge the interests of multinational corporations, and to place human needs before economic efficiency and profit-making;
- more democratic and participative forms of social organization;
- greater international cooperation instead of competition;
- an increasing recognition of the importance of cultural and biological diversity, and locally adapted solutions for sustainable development;
- empowerment of the rural poor through education and training so that they no longer perceive themselves as victims of circumstances;
- greater respect for the environment;
- greater understanding of the human place in nature, and of how our collective activities and individual consumption patterns impact the environment; and
- a merging of sophisticated, information-age technologies with traditional knowledge.

The Chinese curse, 'may you live in interesting times', seems to apply to us today. We live in very interesting times, and while it is easy to see the curse in all of this, there is also perhaps a blessing as well. The blessing is that those of us who are alive today have the opportunity to contribute to the making of an entirely different form of social and economic organization. Fortunately, unlike medieval Europeans at the time of the plague, we are not beholden to forces of nature over which we have little control or understanding. We have a self-awareness and an ability to choose alternative paths that can lead us away from calamity. If we fail to do so, the risk is very great. But if we succeed, future generations will thank us for using our creativity to bring about a more equitable global society that respects the environment while meeting human needs.

Notes

1. The author gratefully acknowledges the financial support of the US Agency for International Development through the University of Michigan Population–Environment Fellows Program. The opinions expressed in this chapter are those of the author, and do not necessarily reflect the views of the aforementioned institutions.
2. IUCN–the World Conservation Union, World Wide Fund for Nature and United Nations Environment Programme, *Caring for the Earth: A Strategy for Sustainable Living* (Gland, Switzerland: IUCN, 1991), 43.
3. A. de Sherbinin, 'World Population Growth and U.S. National Security', *Environmental Change and Security Project Report* 1 (Spring 1995): 24–39.
4. See, for example, the President's Council on Sustainable Development, *Population and Consumption Task Force Report* (Washington, DC: PCSD, 1996), 33.
5. Population Reference Bureau (PRB), *World Population Estimates and Projections by Single Years: 1750–2100* (mimeo, 1995); World Bank, *World Development Report 1992: Development and the Environment* (New York: Oxford University Press, 1992), 32–3.
6. Emulation often involves local variations. For more on this, see R. Wilk, 'Emulation and Global Consumerism,' in P. Stern, T. Dietz, V. Ruttan, R. Socolow, and J. Sweeny (eds), *Environmentally Significant Consumption: Research Directions* (Washington, DC: National Academy Press, 1997), 110–115.
7. J. Tinbergen and R. Hueting, 'GNP and Market Prices: Wrong Signals for Sustainable Economic Success that Mask Environmental Destruction', in *Population, Technology and Lifestyle*, eds R. Goodland, H. Daly and S. El Serafy (Washington, DC: Island Press, 1992), 52–62.
8. See P. Stern, 'Toward a Working Definition of Consumption for Environmental Research and Policy,' in P. Stern, T. Dietz, V. Ruttan, R. Socolow, and J. Sweeny (eds), *Environmentally Significant Consumption: Research Directions* (Washington, DC: National Academy Press, 1997), 12–25.

9. For more on this topic, see D. Korten, *When Corporations Rule the World* (Hartford, Conn.: Kumarian Press, 1995); and H. Daly, 'The Perils of Free Trade', *Scientific American* (November 1993): 24–9.
10. American Automobile Manufacturers Association (AAMA), *World Motor Vehicle Data 1996* (Washington, DC: AAMA, 1996), 2–15.
11. J. Mackenzie and M.P. Walsh, *Driving Forces: Motor Vehicle Trends and their Implications for Global Warming, Energy Strategies and Transportation Planning* (Washington, DC: World Resources Institute, 1990), 7.
12. Data on fossil fuel consumption from *The Statistical Abstract of the United States 1996* (Washington, DC: Government Printing Office, 1996), table 917, 580.
13. M. van Brakel and M. Buitenkamp, *Sustainable Netherlands: A Perspective for Changing Northern Lifestyles* (Amsterdam: Friends of the Earth, 1992); and A. Hittle, *The Dutch Challenge: A Look at How the United States' Consumption Must Change to Achieve Global Sustainability* (Washington, D.C.: Friends of the Earth, 1994).
14. D. Korten, 'Civic Engagement in Creating Future Cities', *Environment and Urbanization* 8 (April 1996): 37.
15. 'A Survey on Living with the Car', *The Economist* (22 June 1996): 4.
16. J. Whitelegg, 'India's Roads to Ruin', *New Scientist* (1 February 1997): 51.
17. 'A Survey on Living with the Car', *The Economist*, op. cit.
18. For more on micro-rationality versus macro-irrationality, see S. Boyden and S. Dovers, 'Natural-Resource Consumption and its Environmental Impacts in the Western World: Impacts of Increasing Per Capita Consumption', *Ambio* 21 (February 1992): 63–9.
19. Cornucopians refers broadly to the proponents of free trade and analysts at conservative think tanks such as the Cato Institute, American Enterprise Institute and Heritage Foundation, among others.
20. W. Kempton and C. Payne, 'Cultural and Social Evolutionary Determinants of Consumption', paper presented to the National Academy of Sciences workshop on the Global Environmental Impact of Consumption in the United States, Washington, DC, November 1995.
21. Soros writes 'In many parts of the world control of the state is so closely associated with the creation of private wealth that one might speak of robber capitalism, or the "gangster state", as a new threat to the open society', G. Soros, 'The Capitalist Threat', *The Atlantic Monthly* (February 1997): 55.
22. Survey research and focus groups conducted by the Pew Charitable Trusts' Global Stewardship Initiative and the Merck Family Fund have found that Americans would be willing to trade consumer goods for reduced stress and more free time.
23. K. Grimsley, 'U.S. Corporations Look for Incentives to Entice Low-Wage Workers to Stay', *International Herald Tribune* (24 March 1997): 7.
24. M. Castells, 'High Technology and the International Division of Labour', *Labour and Society* 14 (1989); cited in 'The City From Here', *Environment and Urbanization* 8 (October 1996): 105.

25. See, for example, G. Martine, 'Population/Environment Relations and International Security: The Impacts of Economic Globalization', paper prepared for the Third Conference on Environmental Security, 31 May–4 June 1994, Tufts University, Somerville, Massachusetts.
26. See Korten, *When Corporations Rule*, 49–50, note 9 above; and National Labor Committee, 'Sweatshop Development', in *The Haiti Files: Decoding the Crisis*, ed. J. Ridgeway (Washington, DC: Essential Books, 1994), 89–112.
27. W. Lutz, 'Population and Biodiversity: A Commentary', in *Human Population, Biodiversity and Protected Areas: Science and Policy Issues*, ed. V. Dompka (Washington, DC: American Association for the Advancement of Science, 1996), 229–42.
28. *Population Summit of the World's Scientific Academies: A Joint Statement by 58 of the World's Scientific Academies* (Washington, DC: National Academy of Sciences Press, 1994).
29. R. Roy, *Experimenting with Truth: The Fusion of Religion with Technology, Needed for Humanity's Survival* (New York: Pergamon Press, 1981), 26–9.
30. S. Moore, 'The Coming Age of Abundance', in *The True State of the Planet*, ed. R. Baily (New York: Free Press, 1995), 112–3.
31. C. Lynch, 'Searching the Internet', *Scientific American* (March 1997): 45.
32. E.F. Shumacher, 'The Age of Plenty: A Christian View', in *Valuing the Earth: Economics, Ecology, Ethics*, eds H. Daly and K. Townsend (Cambridge, Massachusetts: MIT Press, 1993), 161.
33. See, for example, Martine, 'Population/Environment Relations', note 25 above.
34. P. Harrison, *The Third Revolution* (London: Penguin Books, 1992).
35. See Korten, 'Civic Engagement', 35–50, note 14 above.

16 Population, Land-Use and the Environment in a West African Savanna Ecosystem: An Approach to Sustainable Land-Use on Community Lands in Northern Ghana[1]

Gottfried Tenkorang Agyepong,
Edwin A. Gyasi and John S. Nabila,
with Sosthenes K. Kufogbe

The West African savannas comprise east–west vegetation zones differentiated in terms of structure, composition and physiognomy as the result of a rainfall gradient that decreases from south to north. Savanna-forest transition zones adjoin high forest in the south, while arid and semi-arid conditions prevail in the Sahel at the northern margins bordering the Sahara Desert (see Figure 16.1). The high forest to the south and the Sahel to the north have received more attention than the intervening savannas in recent years with respect to study for conservation and development efforts.[2] This is because of concerns about biodiversity, the availability of wood resources and the more favourable moisture conditions for plant growth in the forest regions, and because of the threat of desert encroachment from the north in the Sahel zone. The relatively few studies of humid savanna environments do indicate that they have been subjected to continuing environmental stresses that threaten the future social and economic security of the peoples of the savannas.[3]

The savanna environment is by nature fragile, with relatively low carrying capacity[4] due to poor soils (Plinthic ferrasol), soil moist-

Figure 16.1   Savanna forests and forest transition zones

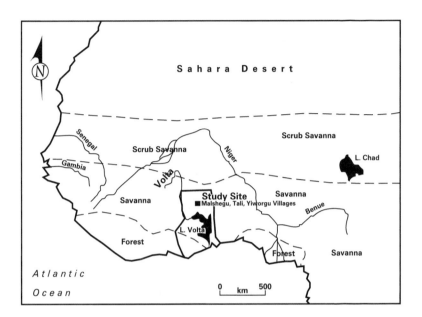

ure deficits and low biomass production.[5] The frequent failure of
development projects in savanna environments worldwide[6] suggests
that model solutions to the human-induced environmental and production problems of the savannas are not available. Nelson suggests
that progress will depend upon small pilot projects, community
experimentation and within-country experience, and that more
progress is likely through the use of incentives that promote spontaneous responses among local communities.[7] It may be added that
the most progress will be made if indigenous knowledge and practice are harnessed.

The above suggestions provide useful hypotheses that should be
combined and tested in the field to identify sustainable development interventions in the savannas, but the design and implementation of small, pilot, sustainable interventions requires baseline
information on both the biophysical and human conditions. The
Cooperative Integrated Project on Savanna Ecosystems in Ghana
(CIPSEG), implemented from 1992 to 1996, collected this type of
information. The work reported in this chapter sought to establish
the baseline for the spatial context of the project (other studies

addressed botany, biodiversity and wildlife, agriculture, animal husbandry and culture and traditions), and involved testing one model experimental design for the guinea savanna environment in northern Ghana based on the practices of 'sacred grove' communities.

Grove communities are defined in terms of their common cultural affinity to a sacred grove, which is an area of original vegetation traditionally preserved through local taboos and sanctions that express ecological and spiritual values. The groves vary in spatial extent from a single tree to a few hectares (ha) of climax or near climax vegetation in areas of otherwise extensively converted or modified vegetation. The ecological values that preserve these groves are founded on the fact that the groves are often located in areas where they protect spring sources or hilltops from erosion. The spiritual values derive from the fact that the groves are believed to accommodate personified natural phenomena such as hills, streams and special plant and animal species or human spirits. One idea tested by CIPSEG is the possibility of using the traditional values embodied in the grove as a vehicle for the introduction of scientific conservation and management of the community land. To do this, three sacred grove communities in the CIPSEG project area were selected, where the socioeconomic conditions of the population and their land-use patterns were representative of those in the savanna areas of Ghana and of West Africa more generally. This study describes the outcomes of the investigations in these three communities.

THE GROVE COMMUNITIES

The three communities studied are Malshegu and Tali in the Tolon-Kumbungu District, and Yiworgu in the Salvelugu-Nanton District of Northern Ghana (see Figure 16.1). A total of 31 closely-spaced villages were identified within these three communities, generally separated by distances of 1.5 to 2.5 kilometers (km) (see Figure 16.2) and linked by numerous footpaths scattered throughout the landscape.

The People

The individual villages are generally small, with populations of less than 500 people, and consisting of clusters of compounds. The

Figure 16.2   Distribution of villages in the Malshegu community area

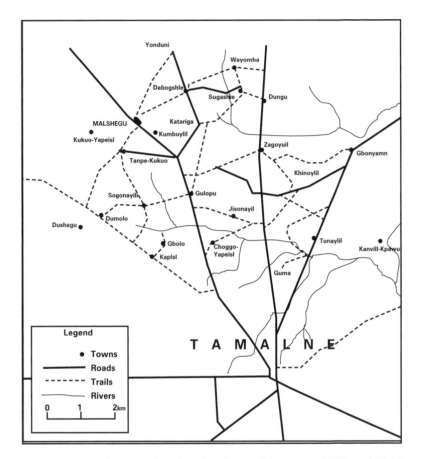

number of settlements has hardly changed between 1960 and 1993, but according to the Ghana census, population has increased in most communities between 1960 and 1984, in some cases by as much as 235 per cent.

Evidence from the survey suggests that the population increases arose from increased births rather than from in-migration; out-migration has been minimal. It is difficult to provide precise population density figures for the study areas because exact aerial delimitations of community lands have not been obtained, but estimates range from nine people per km^2 for the Yiworgu grove community area, to 64 in Tali and 96 in Malshegu.

Men in these communities are polygamous and often have two or more wives. Family sizes are large, with 6 to 15 persons per family, and there is generally a strong desire to have more children. Education and health facilities and water supply are generally inadequate, though primary health care and birth attendance have recently improved. Health problems are low among adults, at less than 30 per cent, but affect as many as 89 per cent of the children, particularly children of weaning age. Nutrition also poses problems for the majority of households. About 62 per cent of them experience frequent food shortages in spite of the large variety of food crops cultivated by the people, indicating widespread food insecurity. Food shortages occur mostly during the farming season, which starts in April, when meeting energy requirements is critical for farm work, and they become most acute just before harvest. Food is often purchased during this period with cash obtained from animal sales.

Other important characteristics observed relate to employment, income and poverty. Employment is understood here to mean any human activity that provides a reasonable source of livelihood for a person, irrespective of the proportion of the person's time occupied or the level of productivity. Based on this definition, there is nearly full employment among the people, with about 98 per cent possessing some form of livelihood. Of these, about 87 per cent are farmers, of whom 45 per cent have secondary occupations. Fibre rope-making is the most important secondary occupation, while mat-making and the rearing of domestic animals such as sheep, goats and fowl are also important supplementary occupations. The fact that about 74 per cent of employed persons have been in their occupations since childhood suggests that occupational mobility is quite limited.

It is difficult to determine the actual monetary income levels of the predominantly rural dwellers in the grove community areas. However, the predominance of essentially subsistence agriculture supplemented by widespread gathering of wild produce indicates low levels of income. The reported inability of many farmers to buy chemical fertilizers and other agro-chemicals to improve crop yields provides additional evidence of low income levels. Land and buildings, in the form of huts clustered in compounds, are the most common properties owned by the people, while ownership of businesses is insignificant.

The Land

Climate is an important factor in the Guinea savanna environment of northern Ghana. The mean annual rainfall is between 1000 and 1200 millimeters (mm), falling in one rainy season from May to November. Sandy soils overlying plinthic materials, together with high evapotranspiration rates, result in widespread soil moisture deficits during the dry season. The climax vegetation consists of a ground flora of perennial grasses less than 1.5 meters (m) high, interspersed with generally fire-resistant, deciduous, broad-leaved trees. In the most luxuriant areas the trees show varying completeness of canopy. However, the climax vegetation hardly exists anywhere, having been modified by annual bush-burning and grazing.[8]

The proportion of uncultivated land that is suitable for agricultural use is quite low in the study area, indicating a high level of pressure on land resources. Estimates from aerial photographs and ground traverses indicate that the proportion of uncultivated land in the grove areas has decreased from 60 per cent in 1960, to about 17 per cent in 1993, and large proportions of this land are either unsuitable or marginal for cultivation due to iron pans, flood hazards, poor soils and rock outcrops. The general environment is degraded, with the woody cover extensively removed and soils visibly impoverished.

Land ownership and tenure in the community areas are rooted in traditional common property systems. Land-tenure status can actually be sub-divided into land tenure and tree tenure.[9] Land is collectively owned by kinship groups, with custody and control vested in either *Tindamba*, the spiritual head of the community, or in the secular head who is also the traditional chief. Members of the community have the right of free access or use of the land for farming, grazing, gathering and housing. This right is exercised individually or, more commonly, on a family basis, without prejudice to the principle of common ownership. The right of usage established over a particular parcel of land is heritable patrilinearly, but traditionally the land is not sold. However, owing to the growing demand for land, particularly for housing, land markets are developing.

Tree tenure concerns the rules and regulations that govern the ownership and access to trees of economic or other value. According to customary practices, the right to harvest, collect, fell and otherwise use trees generally rests with the custodian, who is the chief or the *Tindamba*, except in the case of firewood and certain fruits from naturally propagated trees, which may be harvested by anybody from the community. In some cases fruits may be shared between the custodian or the chief and the farmer who works on the land.

The customary tenure arrangements ensure that land is available for all members of the community who need it for subsistence farming and for housing, particularly when the population is small. However, with increasing population, monetization of the economy, and extremely limited alternative employment opportunities in the immediate and surrounding areas, the systems are currently undergoing changes, including sale of land resulting from growing pressures on farm and grazing land. This is already a problem in the Malshegu grove community area, where the effect of the regional capital is gradually extending outward.

The land-use systems prevalent in the study areas are shown in Figures 16.2 and 16.3. The indigenous system of bush-fallow agriculture remains the dominant system of crop production. This system is characterized by mixed cropping of vegetables and traditional staples such as sorghum, millet and yams, and the use of varying periods of natural fallow and regrowth for the restoration of soil fertility. These crops are produced primarily for subsistence, with any surpluses sold on the market at Tamale, the regional administrative centre. Livestock rearing is done mainly using a free-range system. The most important animals – cattle, sheep, goats and fowl – are traditionally kept primarily for social occasions such as marriage, but also as security to sell when cash is needed to purchase food after crops fail, or to cover medical expenses. Gathering wild produce for food, medicines, fuel and construction is an equally important practice.

Ground-traverse observations in the grove community area reveal that the basic pattern of land-use common in all of the villages consists of approximately concentric zones of land-use differentiated by the distance from the built-up area, the crop emphasis, levels of fertilization, crop protection and general care (see Table 16.1). Immediately adjoining the family clusters of huts and the built-up area are the gardens or 'compound farm' plots, which are cropped annually to staples and vegetables and fertilized with compost from household refuse and dung from kraals (animal pens), which may also be located in this area.[10] Every compound owns a garden in this zone. The compound farm zone generally extends up to about 300 m from the residential area. Fencing

Figure 16.3   Land-use and cover in the Malshegu community area

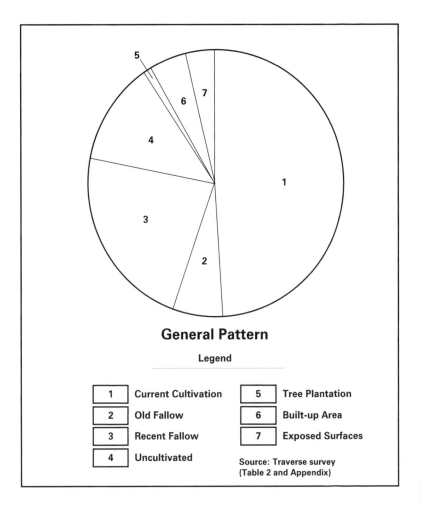

General Pattern

Legend

1	Current Cultivation	5	Tree Plantation
2	Old Fallow	6	Built-up Area
3	Recent Fallow	7	Exposed Surfaces
4	Uncultivated		

Source: Traverse survey
(Table 2 and Appendix)

with corn stalks and sticks serves to protect crops from animals, while proximity to the village facilitates intensive care of these plots.

Bush-fallow farming is practiced in the next zone, where staple crops are grown, and rice may also be cultivated away from the village in the valley bottoms. Soil fertility is maintained through the practice of bush fallowing, which requires that the farmer move on to cultivate another piece of land after one to three years of cultivation, while the previous one lies fallow. The cycle for the

Table 16.1   Spatial structure of community land-use in Katariga village

Garden or compound farm zone (up to 200–300m)	Bush-fallow farm zone (750 m)	Uncultivated land or other village land
• Gardens Plots	• Combination of fallow and extensive current cultivation	• Mainly intrazonal within bush farm zone
• Manuring practiced		
• Gardens penetrate built-up area	• Little area under recent fallow	• Extensive grazing
• Temporary fences around gardens	• Few old fallows	• Wood gathering
• Reduced tree cover	• No fences/hedges	• Brush burning
• Tethered animals fed with lopped branches	• Mixed cropping (maize, beans, groundnuts)	
• Grazing in dry season	• Not fertilized	
• Kraals	• Grazing restricted to follow areas	
• Hand ploughing with hoe	• Selected tree cover (widely spaced eg. shea and dawadawa trees)	
• Garden ends in bush farms	• Bullock and hand-ploughing	

Source: G.T. Ageypong; 'Population and Land Use in a West African Savanna Ecosystem: A Case Study of Environment Management and Conservation In the Tolon-Kumbungu District of Northern Ghana', paper presented to ICSE Conference.

recultivation of the fallow land, and thus the level of soil-fertility restoration, depends upon the degree of pressure on or demand for land.[11] Full restoration of fertility requires fallow periods of up to 15 years, but the demand for land in the study area is now so high that land may be fallowed for only one or two years, a period which is inadequate for soil-fertility restoration.

The extent of the bush-fallow zone varies and may depend on land pressure near the village, the development of roads and the availability of means of transport such as bicycles for daily commuting to the farm. While the limits of the compound gardens or farms are determined by the need for intensive management, the limits of the bush-fallow farm seem to be determined by how much food is needed to supplement production in the compound zone. In most of the villages, including Malshegu, almost all suitable village farm land is already cultivated.

Changes in the Spatial Structure of Land-Use

The basic structure of village community land-use has remained essentially unchanged between 1960 and 1993, but some modifications have taken place, most noticeably in the bush-fallow zone. These changes have taken the form of the extension of this zone farther away from the village due to both the increased availability of road transport such as bicycles, and to the introduction of cash crops such as rice in the relatively distant valley bottoms. Nearer to the villages, farmers may use chemical fertilizers to make up for the shortened fallow if they can afford it, though few can. The farmers have also found, however, that the initially favourable response of soils to fertilizer does not last, a problem which is officially attributed to inappropriate applications. Other modifications include the introduction of new crops, especially cassava which is valued for its efficient utilization of soil moisture, and swamp rice which is grown as a commercial crop.

In the past, the bush-fallow zone has been the zone of agricultural extensification to accommodate growing village populations. However, in many villages these zones now extend to the limits of village lands, and now border the similarly expanding bush-fallow zones of neighbouring villages. Thus, the region is now dominated by large, continuous areas of cultivated land with increasingly impoverished soils and little woody cover or uncultivated land remaining.

One of the difficult problems of land-resource management in the vast area of the savannas is the relative importance of crop agriculture and free-range livestock rearing. The practice of rearing livestock is deeply rooted in the savanna cultures;[12] in the drier savanna environments in the Sudan and the Sahel, the nomadic cultures emphasize animal rearing. Although in the more humid savannas of the grove communities agriculture is certainly more important in terms of the effort expended on it, livestock is also an important element of local livelihoods. However, in both zones it is becoming increasingly difficult to support the growing numbers of people and livestock, posing a serious danger of severe land degradation. Sustainable development of the two regions may therefore require that animal numbers be limited and their quality

Table 16.2 Environmental processes, indicators and incidence in the study areas

Process	Indicator	Frequency	Incidence
Burning	Fire scars, burnt stumps and tree bark	common common	Uncultivated and long fallow areas
Free grazing	Scattered dung	common	Fallow lands
	Animal trails	common	Uncultivated lands
	Kraals	common	Tether points
	Grazing animals	common	Cultivated fields
	Gully erosion	common	Village compounds
	Sheet erosion	common	Paths
		common	Upper slope fields
		common	Stream channels
Tree-lopping	Stumped branches	common	Farm fields
Wood-cutting	Wooden structures	common	Uncultivated areas
	Implements	common	Fallow lands
	Fuelwood	common	
Quarrying and burrowing	Sand and gravel pits	common	Within settlements
	Earth structures	common	Along roads
			Construction sites
Urban encroachment	Peri-urban development	uncommon	Northen parts of Malshegu Community Area

Source: Agyepong, G., T. Nabila, J.S.; Gyasi, E.A. and Kufogbe, S.K. (1993) Aspects of the wider spatial context of the UNESCO Co-operative Integrated Project on Savanna Ecosystems in Ghana. Report of a Baseline Study. Tamale (UNESCO-CIPSEG).

improved, but it is not yet clear how this can be done, especially in an environment where biomass production is so low.

ENVIRONMENTAL IMPACTS OF THESE LAND-USE PRACTICES

The spatial structure of the traditional land-use system leads to a concentration of human activity and animals near to settlements, thus putting considerable pressure on the land immediately surrounding the villages. Meanwhile, the reduction in fallow periods in the region seems to be a major factor in the decline of the physical and chemical condition of the soils. Bush-burning, free grazing, soil erosion and tree removal are common practices and processes that also degrade the land (see Table 16.2). It has not been possible to study these processes in quantitative terms, but qualitative observations have been made during all of the field traverses. Evidence of bush fires, a key cause of land degradation, was widespread. Fire scars were obvious in the uncultivated areas, including the preserved sacred groves. In the past, bush fires have swept through all of the sacred groves, and have fragmented them or led to their contraction at the edges due to the increased chances of tree death and removal after the fires. Even though fire may be used as a management tool in savanna ecosystems,[13] in the present environment the indications are that bush-burning has been uncontrolled.

Damage caused by over-grazing was also widespread due to the ubiquitous presence of large and small ruminants on open range land. While goats feed in the area immediately surrounding the villages, sheep and cattle are taken further afield for grazing. They are usually shepherded by young boys who protect the animals from theft and prevent them from going into farms during the cropping season from May to November. Animals are left to roam freely during the dry season, at which time all crop residues are eaten and the land is frequently over-grazed. Where animals are kraaled, dung is collected and used as manure in the garden zone. Apart from fires and crop cultivation, animals appear to be the most important factor contributing to the elimination of vegetative cover in the grove community lands.

One of the most damaging effects of over-grazing and burning is sheet and gully erosion, and poor cultivation practices further contribute to these processes. Sheet erosion is a common cause of soil loss, which leads to poor crop conditions and low yields. Gully erosion is more localized. Gullies are generally small and located on cultivated fields, although deeper and wider gullies do occur along main lines of drainage. Quarrying and burrowing for construction materials, another widespread practice, also contribute to this problem, often creating conditions for the initiation of erosion. Burrow pits, which can cover areas of as much as 100m or more in diameter, and may be more than one metre deep, have encroached on both the sacred groves and the open cultivated lands.

The most readily perceived effect of current land-use practices outside of the groves themselves is the loss of woody cover over extensive areas. At present, the average tree canopy cover is only 10 to 12 per cent, made up of selectively preserved indigenous trees in the fields, and introduced species such as mango and neem in and around the villages. Loss of tree cover further contributes to erosion and gully formation, as well as to low soil-moisture content.

The net environmental impacts of these land-use practices have included a decline in the productivity per unit of land and per unit of farmer effort. Marginal lands are increasingly being cultivated, including land at the edge of rock outcrops, and gravelly soils. Food security has also been seriously impaired.

LOCAL PERCEPTIONS OF ENVIRONMENTAL PROBLEMS

Eight of the 31 villages participated in focus-group discussions to explore people's perceptions of their environment and the problems they experience in relation to it, especially those arising from changes in the biophysical environment. In each village, respondents were divided into three groups composed of elderly males, young males and women. The analysis of the transcribed responses from these discussions show only minor variations with respect to local people's conceptions of, perceptions about, attitudes towards, and expectations of their land and environment and their relationship to it. While abstract concepts such as 'environment', 'landscape', and 'watershed' were not very meaningful to people and so could not form a useful basis for discussion, more immediate and concrete issues and problems such as the loss of surface cover and wildlife, decreasing yields and declining rainfall and recurrent drought are clearly understood and widely recognized problems for the people in these villages.

Perceived Environmental Changes

The changes perceived by the elderly men, the women and the young men (15 to 30 years old) are shown in Table 16.3. Not surprisingly, the extreme loss of woody cover which now leaves an estimated tree density of only about seven indigenous trees per hectare, was well noted by the elderly people in most of the villages. Lack of cover is most obvious in the dry season when crops and grasses dry up and are harvested or grazed. The trees that remain are almost all selectively preserved, or have been planted for their economic value, such as the shea tree (*Butyrospermum parkii*) or the dawadawa (*Parkia felicoides*).

Reduced rainfall, declining soil fertility and loss of wildlife were

Table 16.3   Perceptions of environmental changes

Element of change perceived	1 E	1 W	1 Y	2 E	2 W	2 Y	3 E	3 W	3 Y	4 E	4 W	4 Y	5 E	5 W	5 Y	6 E	6 W	6 Y	7 E	7 W	7 Y	8 E	8 W	8 Y
Loss of land cover	◆			◆	◆	◆	◆	◆	◆	◆	◆		◆			◆			◆				◆	
Reduced rainfall amount		◆		◆	◆	◆	◆	◆	◆	◆	◆		◆	◆		◆	◆		◆	◆			◆	
Unreliable rains											◆						◆				◆			
Drying water bodies															◆									
Droughts												◆										◆		
Declined soil fertility	◆				◆	◆	◆	◆	◆		◆			◆	◆		◆	◆		◆	◆		◆	◆
Loss of wildlife	◆			◆	◆	◆	◆	◆	◆				◆	◆		◆			◆			◆	◆	
Short rainy season										◆			◆			◆			◆					

1 – Katariga-Kumbuyili; 2 – Zaguyuli; 3 – Gurugu; 4 – Chogo Yepalsi; 5 – Kpiendua; 6 – Yepalsi; 7 – Chirifoyili; 8 – Dalinbihi. E – Elders; W – Women; Y – Youth.

Source: Agyepong, G., T. Nabila, J.S.; Gyasi, E.A. and Kufogbe, S.K. (1993) Aspects of the wider spatial context of the UNESCO Co-operative Integrated Project on Savanna Ecosystems in Ghana. Report of a Baseline Study. Tamale (UNESCO-CIPSEG).

also identified by most of the groups as key environmental problems. These are environmental issues that affect the lives of the people directly, and they are the subjects of local discussion and religious observances. One of the functions of the land priest, the *Tindamba*, for example, is to invoke the gods to bring rains when they fail. Average annual rainfall data from a nearby agricultural research station actually show that over the past 27 years there have been as many years with below average rainfall as above average. Persistent reports of decreased rainfall may therefore indicate that the perception of change arises from alteration in the seasonal distribution, rather than changes in the mean annual totals. Large variations in the onset and cessation of rain over an extended period may have similar effects on agriculture as reduced total rainfall.

The loss of wildlife in the region strikes any visitor to the community lands, and local residents also recognize this and identify it as a key problem. The only wild animals encountered during the field traverses were a few small birds and a single alligator, which may have survived only because it is taboo to kill them. Where there was once an abundance of wild animals such as lions and other predatory animals and game, the region has now been depleted of almost all large animals as a result of the destruction of vegetation and wildlife habitat.

PERCEIVED FACTORS OF CHANGE

Local perceptions about the factors that cause problems of environmental degradation are indicated in Table 16.4. There is no doubt that factors such as poverty and the failure to adopt improved technology are important in the explanation of environmental deterioration in the grove community lands and in the guinea savannas in Ghana generally. Population growth, however, is commonly held to be the key factor, even by the local people. Population has more than doubled in almost all of the communities between 1960 and 1984; only three small villages actually decreased in size. The absolute population numbers are not striking, but as indicated earlier the densities are relatively high. Visible evidence of stress in the environment, including loss of vegetative cover, soil erosion and poor crop conditions correspond well with local population densities, Malshegu being the most severely affected.

Table 16.4 Local perceptions about factors that cause problems of environmental degradation

Element of change perceived	1 E	1 W	1 Y	2 E	2 W	2 Y	3 E	3 W	3 Y	4 E	4 W	4 Y	5 E	5 W	5 Y	6 E	6 W	6 Y	7 E	7 W	7 Y	8 E	8 W	8 Y
Population growth	◆	◆		◆	◆		◆	◆		◆	◆		◆			◆			◆					
Decreasing rainfall																				◆				
Deforestation	◆	◆		◆			◆				◆			◆			◆							
Hunting	◆	◆			◆			◆			◆			◆	◆		◆			◆			◆	
Settlement encroachment	◆			◆	◆		◆			◆														
Inappropriate technology	◆						◆							◆			◆							
Non-traditional practices	◆							◆						◆			◆					◆		
Road/other construction	◆	◆		◆	◆		◆				◆						◆							
Soil exhaustion		◆			◆		◆			◆			◆			◆								

1 – Katariga-Kumbuyili; 2 – Zaguyuli; 3 – Gurugu; 4 – Chogo Yepalsi; 5 – Kpiendua; 6 – Yepalsi; 7 – Chirifoyili; 8 – Dalinbihi. E – Elders; W – Women; Y – Youth.

Source: Agyepong, G., T. Nabila, J.S.; Gyasi, E.A. and Kufogbe, S.K. (1993) Aspects of the wider spatial context of the UNESCO Co-operative Integrated Project on Savanna Ecosystems in Ghana. Report of a Baseline Study. Tamale (UNESCO-CIPSEG).

It seems inadequate, however, to explain the conditions of the community lands only in terms of population densities. In addition to growing population, local people also consider the disregard of traditional practices, including religious observances, to be a cause of environmental deterioration. Other important factors may also include the inability of the traditional bush fallow method of cultivation to cope with the increase in population; the traditional spatial land-use structure, which results in a concentration of people and animals in relatively small spaces (done for security reasons in the past); the excessive dependence of the people on the local environment for most of their biophysical needs for food, wood and fibre; and the use of fire as an agricultural tool.

SUSTAINABLE MANAGEMENT

The information available here suggests that the objective of sustainably managing resources and the environment can be achieved if:

1. woody cover is rejuvenated and maintained to provide wood, food and energy resources;
2. animal husbandry is improved through the increase of fodder resources; and
3. crop production is stabilized.

In the past, environmental and socioeconomic problems have led to government interventions based on population movements and resettlement,[14] and the need for improvements in farm management and conservation techniques has also long been acknowledged.[15] The integrated rural development approach was the basis of such efforts in Ghana during the 1980s.[16] The Land and Water Management Project, for example, involved the preparation and implementation of village land development plans, including adaptive trials and demonstrations, and community-implemented conservation measures; this project is currently in a pilot phase in both forest and savanna environments in Ghana.[17]

CIPSEG is a research and development intervention project. The unique aspect of this project lies in its conscious effort to understand and to harness traditional knowledge, practice and experience in environmental conservation as embodied in the concept of the 'sacred grove', a common cultural phenomenon in both savanna and forest environments. The approach to sustainable development envisaged in CIPSEG seeks to use the sacred grove as the focus for scientific conservation interventions in the associated community lands, combining the traditional concepts and practices of the sacred grove and the zonation of land-use around the house compound areas. The conceptual framework adopted is that of the Biosphere Reserve, 'a biotic province that harbours sample communities of species'. The spatial structure of reserves often exhibits a zonal pattern that provides different levels of protection and management. The design of the sustainable development intervention in the sacred grove community lands envisages the conservation management of the community land as a single biosphere reserve system with a fully protected grove core, a buffer zone of controlled development and exploitation of resources, and an outer zone of cooperative and assisted planning for improved agricultural and related land-uses (see Figure 16.4).

The grove core, which consists of a pro-climax savanna woodland, has educational and demonstration value for biodiversity, recreation and tradition. It is protected from fire and exploitation to allow natural regeneration. The buffer zone may include portions of the original grove area that have been encroached upon for farming or grazing; it is communally managed for reforestation, wood harvesting and controlled grazing. The community is assisted in delimiting this zone to avoid conflict. The outer area constitutes the settlement and farming zone. It is envisaged as an area of improved land-use practices, including the use of manure, the development of fodder banks, woodlots, tree nurseries and live fencing, and the practice of agroforestry.

These interventions have been initiated in the three grove community areas in this study, involving a wide spectrum of socioeconomic and biophysical data acquisition, field demonstrations, and the encouragement and development of local initiatives and capacity.[18] The experience is that if nothing is imposed, and if the people are made to take control and are motivated, there is a good chance of making progress together. The people have learned to develop tree nurseries, individual woodlots, agroforestry practices and fodder banks to provide for animals during the dry season. They have also been exposed to ideas of biodiversity, conservation and simple field observation of environmental phenomena.[19] The communities now have a concept that relates not only to their spiritual perceptions, but also to their physical well-being. The sacred grove as a sociocultural and educational vehicle for the sustainable development of savanna

Figure 16.4   Grove management model

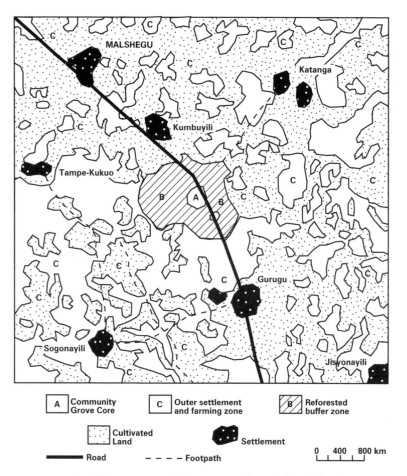

resources in Ghana may well prove to be viable. The long-term viability will depend, however, on continued support for building local capacity and confidence.

Notes

1. This text is derived from the participation in, and report of the authors to, the Cooperative Integrated Project on Savanna Ecosystems in Ghana (CIPSEG), financed by the German Federal Ministry of

Economic Cooperation and Development during the period 1992–95. A great deal of insight was gained from the many discussions that took place among the scientists who took part in the project at various project meetings.

2. See, for example, J.B. Hall and M.D. Swaine, *Distribution and Ecology of Vascular Plants in a Tropical Rainforest: Forest Vegetation in Ghana* (Netherlands: Dr W. Junk Publishers, 1981); J.B. Hall and M.D. Swaine, 'Classification and Ecology of Closed Canopy Forest in Ghana', *Journal of Ecology* 64 (1976); and K. Cleaver, M. Monasinghe, M. Dyson, N. Egli, A. Peuker and F. Wencelius, *Conservation of West and Central African Rainforests* 1 (1992): 353.

3. See, for example, UNSO, *Assessment of the Problem of Desertification and Review of Ongoing and Proposed Activities to Implement the Plan of Action to Combat Desertification in Ghana* (Accra: EPC, 1986); and EPC/UNSO, *A Socio-Economic Survey of the Upper East Region with Reference to the Drought and Desertification Control in Ghana* (Accra: EPC, 1993).

4. O.T. Solbrig, 'Ecological Constraints to Savanna Land Use', in *The World's Savannas: Economic Driving Forces, Ecological Constraints and Policy Options for Sustainable Land Use*, eds M.D. Young and O.T. Solbrig, Man and Biosphere Series no. 12 (Paris and Camforth: UNESCO and Parthenon Publishing, 1993).

5. M.D. Young and O.T. Solbrig, 'Savanna Management for Ecological Sustainability, Economic and Social Equity', *MAB Digest* 13 (1992).

6. See, for example, Young and Solbrig, 'Savanna Management', *ibid.*; R. Chambers, *Settlement Schemes in Tropical Africa: A Study of Organisation and Development* (London: Routledge & Kegan Paul, 1969); G.T. Agyepong, 'Developing Institutional Infrastructure for Handling Conservation Problems in West Africa', technical papers from a consultation on Promoting Conservation for Sustainable Development in the Sudo-Sahelian Region of Africa, Banjul, The Gambia, 1989.

7. R. Nelson, *Dryland Management: The 'Desertification' Problem*, World Bank Environment Department Working Paper no. 8 (Washington, DC: World Bank, 1988).

8. R. Rose-Innes, *Land and Water Survey of the Upper and Northern Regions of Ghana*, FAO/sp:31/GHA (Rome: FAO, 1963).

9. See for example, C. Oppong, *Growing up in Dagbon* (Accra: Ghana Publishing Corporation, 1973); A. Panin, *Hoe and Bullock Farming Systems in Northern Ghana: A Comparative Socio-Economic Analysis*, Nyankpala Agricultural Research Report no. 1 (1988); R.J.H. Pogucki, *Gold Coast Land Tenure: A Survey of Land Tenure in Customary Law of the Protectorate of the Northern Territories* (Accra: Lands Department, 1955); and M. Staniland, *The Lions of Dagbon: Political Change in Northern Ghana* (Cambridge: Cambridge University Press, 1975).

10. J.B. Wills, ed., *Agriculture and Land Use in Ghana* (Oxford: Oxford University Press, 1962).

11. G. Benneh, 'Systems of Agriculture in Tropical Africa', *Economic Geography* 48 (1972).

12. Solbrig, 'Ecological Constraints', notes 4 above.

13. J.M. Ramsay and R. Rose-Innes, 'Some Quantitative Observations on the Effects of Fire on the Guinea Savanna Vegetation of Northern Ghana over the Period of Eleven Years', *African Soils* VIII (1963): 41–85.
14. See, for example, T.E. Hilton, 'Frafra Resettlement and the Population Problem in Zuarungu', *Universitas* 3 (1959): 144–6; T.A.M. Nash, 'The Anachu Settlement Scheme', *Farm and Forest* II (1941); J. Dundas, 'Illorin Rural Planning Scheme', *Farm and Forest* III (1942); Chambers, *Settlement Schemes*, note 6 above; and G. Brown, 'Experiment in Co-Operative Farming at Daudawa in Northern Nigeria', *Empire Cotton Growing Review* VIII (1930).
15. See, for example, J.N. Pittman, *Land Planning, Soil and Water Conservation in Northern Ghana: A Review of Progress over the Past 10 Years* (Accra: Department of Agriculture, 1959); Colonial Office, *British African Land Utilization Conference*, Jos, Nigeria, 7–15 November 1949 (London: HMSO, 1949); C.W. Lynn, 'Land Planning, Soil and Water Conservation in the Northern Territories', *Farm and Forest* 7 (1945).
16. Ghana/CIDA/Wardrop-Deloitte, *The Northern Region, Ghana, Volume 1: A Descriptive Overview* (Tamale: NORRIP Technical Unit, 1983).
17. DANIDA/Cowi Consult, *Land and Water Management Project in Ghana*, project document, 1994.
18. E. Telly and A. Fiadjoe, *Environmental Education and Training for Savanna Ecosystem Management* (Tamale: UNESCO–CIPSEG, 1996).
19. UNESCO, *CIPSEG: Project Findings and Recommendations* (Paris: UNESCO, 1986).

Reference

Nye, P.H. and D.J. Greenland (1960) *The Soil Under Shifting Cultivation*, Technical Communication no. 51 (Harpenden: Commonwealth Bureau of Soils).

17 'stay together/learn the flowers/go light': Education in a Time of Environmental Crisis[1]

John Elder

We have entered an era of environmental crisis that requires a new look at our ideas about liberal education. 'For the Children', a poem by Gary Snyder from his Pulitzer Prize-winning collection *Turtle Island*, evokes the challenges faced by those living today:

> The rising hills, the slopes,
> of statistics
> lie before us.
> the steep climb
> of everything, going up,
> up, as we all go down.
>
> In the next century
> or the one beyond that,
> they say,
> are valleys, pastures,
> we can meet there in peace
> if we make it.
>
> To climb these coming crests
> one word to you, to
> you and your children:
>
> *stay together*
> *learn the flowers*
> *go light*[2]

It is rare enough to find a poem containing the word 'statistics'. But this one also, in effect, describes a graph. More specifically, the graph described in Snyder's first stanza represents an exponential curve – the steeply rising slope of our own day, in which human population, consumption of fossil fuels, extinction of species, and accumulation of carbon in the atmosphere all go 'up,/up, as we all go down'. Imagine this graph – a curve defined by the x and y axes that extends out almost horizontally until, rather suddenly, it veers steeply upwards. There are two other implicit lines on the graph as well, which parallel the principal axes. The vertical line is invisible, but definite in location; it is the asymptote, a boundary that the rising curve approaches but never touches, instead growing ever more parallel to it. The upward thrust of the curve as it parallels the asymptote charts the politician's dream of endless growth – a rocket blasting through gravity to glide forever along its own triumphant vector. But there is also a second dotted line on this graph, a horizontal one that represents the limits of the earth's carrying capacity.

The term 'catastrophian' has recently been used to describe those who believe that the earth does in fact have a limited carrying capacity. The economists who use this language refer to themselves as 'cornucopians'. They assert, in effect, that no particular natural resource can be totally exhausted, because increasing scarcity causes prices to rise, which in turn leads consumers to choose cheaper, alternative commodities. What this blithe confidence does not address, however, is where a cheaper, alternative planet can be found.

In their 1994 article 'Natural Resources and an Optimum Human Population', David Pimentel *et al.* describe the rapid spiral into poverty that has already accompanied increasing human populations in some areas, and along with this increasing human misery and a drastic decline in biodiversity worldwide. The US population, with an annual growth rate of 1.1 per cent, is expected to double to over half a billion people within 63 years. The world population, now approaching 6 billion and increasing at a rate of 1.7 per cent annually, will double within 41 years. Even China, the world's most populous nation and one with an official policy of one child per family, continues to grow at a rate of 1.4 per cent per year. Self-styled cornucopians must therefore explain how a decent living situation for human beings and a diverse biosphere can be maintained in a world in which population is projected to reach 8.4 billion people by 2025, and 15 billion people by the year 2100.

This situation is, of course, compounded by the fact that US citizens consume an average of 23 times more goods and services than the average citizen of the Third World, and 53 times more than a Chinese citizen.[4] As people around the world aspire to the levels of consumption that characterize the highly publicized way of life in the US, where will they find a new Third World from which to import the necessary resources, and where will they find a new atmosphere into which to dump their carbon?

Ross Gelbspan's 1995 article in *Harper's*, 'The Heat is On', points to the drastic changes in world climate that have already taken place.[5] He reports on the unusual degree of consensus among the 2500 climate scientists that attended the Intergovernmental Panel on Climate Change. Their report, issued in September 1995, confirmed the role of carbon emissions in promoting the greenhouse effect. This causes global warming, leading to the drastic outcomes vividly described by Bill McKibben in *The End of Nature*.[6]

One of our seniors once referred to her Environmental Studies major at Middlebury College as 'the hundred-thousand-dollar migraine' (though I hope she did not mean it!). While it is true that we must begin by facing up to the gravity of the present situation, the next crucial step is to ask ourselves where we can go from here. Snyder's second stanza offers his own vision, one that transforms the graph described above into something more like a Chinese ink painting, with habitable valleys and pastures rolling away beyond the present's looming brow. One way or another, the curve must ripple back down; the real issue is, in Snyder's words, 'if we make it'. Can we call upon the resources of our artistic, spiritual and scientific traditions to help guide the process and, at least to some extent, mitigate the suffering? Can we find, as Snyder seeks to do in this stanza, a beautiful and sustainable vision, uniting us in a community of effort?

Another way to understand our present challenge – to look steadfastly at the worst while also affirming life and moving ahead in hope – is as a form of grieving. As Elisabeth Kubler-Ross pointed out in her book *On Death and Dying*, grieving is hard work and we often seek to avoid it. Denial, anger and negotiation are three of the means of evasion that she describes, and we can see these tactics practiced all around us. The emphasis upon endless economic 'growth' in the platforms of both major political parties resembles an adolescent's refusal to accept limits. But even American nature-writing can manifest its own form of denial; solitary revelations deep in the woods can serve as a form of escapism, ignoring both the ruin of other landscapes, and the distressing circumstances of many urban dwellers. Edward Hoagland has inveighed against writers he describes as 'mystical transcendentalists', who engage in 'Emersonian optimism', arguing that the time for such serene reveries is past. 'Emerson would be roaring with heartbreak and Thoreau would be raging with grief in [our day]. *Where were you when the world burned? Get mad, for a change, for heaven's sake!'*[7]

Hoagland's claim that the best response now is to get mad is, however, flawed, as is the tendency of a number of writers who have influenced environmental thought to blame the Western tradition for contemporary environmental problems. Lynn White, Jr., for example, in his often-cited essay on 'The Historical Roots of Our Ecologic Crisis', points to the creation stories in *Genesis* and their emphasis on 'dominion'.[8] The problem with such arguments, however, is that they tend to be ahistorical. Taking dominion over and replenishing the earth meant something altogether different for an ancient pastoral people than it does for us today. We are called upon not to trash the Bible, but to generate a serious dialogue about how its teachings relate to modern circumstances and conditions and the present state of our knowledge about population–environment relations.

Laying blame is always a way to avoid finding solutions. Moreover, in the face of environmental disaster the blame must extend to an ever-widening circle that finally encompasses all of humanity. The language of our 1964 wilderness legislation, which includes the terms 'untrammeled' and 'pristine', makes a mistake in defining wild lands in terms of opposition to human presence and activity, leading to a sense that the less of *us* there is, the purer and more valuable *it* will be. Placing blame also abounds in the field of literary studies, where it sometimes seems that the angriest, most extreme voices drown out the rest. The German proverb *'Wer schreit, hat recht'* – the one who yells loudest is right – seems to have gained wide acceptance. But those who harshly reject the Western tradition simply inspire a counter-polemic from those who cherish it as a uniquely valuable canon. Culture should instead be seen as an evolutionary process. A cultural tradition can neither be lightly dispensed with nor accepted as a self-contained and knowable artifact. Rather, as T.S. Eliot has explained in his essay on 'Tradition and the Individual Talent', we must enter into our past, and through our greatest creative efforts both sustain it and make it new.

In attempting to sift through both the literary tradition and contemporary environmental thought, the poetry of William Wordsworth is of special value. For Wordsworth, the meaning of nature is always a personal meaning, yet it is a meaning that becomes apparent only as he fixes his gaze upon, and takes his excursions through, the dynamic, non-human world. Seamus Heaney's new collection of Wordsworth's poetry contains these fragments from the poet's *Alfoxden Notebooks*:

> Why is it we feel
> So little for each other, but for this,
> That we with nature have no sympathy,
> Or with such things as have no power to hold
> Articulate language?
>
> _____
>
> And never for each other shall we feel
> As we may feel, till we have sympathy
> With nature in her forms inanimate,
> With objects such as have no power to hold
> Articulate language. In all forms of things
> There is a mind.[9]

In these two passages, Wordsworth enters into the beauty of 'inanimate' nature through a process of connection and sympathy. In a similar fashion, the most effective response to the environmental crises of the modern era will arise neither from a sense of *noblesse oblige* regarding protection of some static conception of 'nature', nor from self-preservation in its narrow sense. Rather, it will arise from a sense of identification both with 'all forms of things', and with 'each other'.

Finally, just as denial and anger cannot resolve our environmental problems, negotiation – a mode often associated with seeking technological solutions to our pressing environmental problems – is also a form of evasion. There are two main problems with relying on techno-fixes. One is that this reliance tends to be based on naive faith rather than a concrete programme, reminiscent of the Dickens character, Mr Micawber, who was always expecting that 'something will turn up'. The other is that it locates our problems outside of ourselves, when in fact the environmental crisis, like the Western tradition, is finally nothing other than ourselves.

To grapple with these problems effectively rather than avoiding them requires that all of the scientific knowledge available today must be augmented by the insights of other disciplines. This mandate to forge a less compartmentalized mode of education regarding environmental issues is why, for me, teaching in both English and Environmental Studies is less a *'split*-appointment' than an experience of deep connections. Nature-writing arises from a confluence of imaginative literature with the natural sciences. It is by joining forces in this way that we can follow Snyder's lead and 'climb these coming crests'.

One crucial model of such a synthesis in American nature-writing is offered by Aldo Leopold. In 'Thinking Like a Mountain', from his *Sand County Almanac*, Leopold recognizes that the heedlessness of our modern approach to nature is both the problem and the context for new insights. He describes how, as a young man working for the US Forest Service in New Mexico Territory, he shot a wolf and watched it die:

Only the ineducable tyro can fail to sense the presence or absence of wolves, or the fact that mountains have a secret opinion about them.

My own conviction on this score dates from the day I saw a wolf die. We were eating lunch on a high rimrock, at the foot of which a turbulent river elbowed its way. We saw what we thought was a doe fording the torrent, her breast awash in white water. When she climbed the bank toward us and shook out her tail, we realized our error: it was a wolf. A half-dozen others, evidently grown pups, sprang from the willows and all joined in a welcoming melee of wagging tails and playful maulings. What was literally a pile of wolves writhed and tumbled in the center of an open flat at the foot of our rimrock.

In those days we had never heard of passing up a chance to kill a wolf. In a second we were pumping lead into the pack, but with more excitement than accuracy: how to aim a steep downhill shot is always confusing. When our rifles were empty, the old wolf was down, and a pup was dragging a leg into impassable slide-rocks.

We reached the old wolf in time to watch a fierce green fire dying in her eyes. I realized then, and have known ever since, that there was something new to me in those eyes – something known only to her and to the mountain. I was young then, and full of trigger-itch; I thought that because fewer wolves meant more deer, that no wolves would mean hunters' paradise. But after seeing the green fire die, I sensed that neither the wolf nor the mountain agreed with such a view.[10]

Leopold's passage suggests how growth might be possible if we – at times painfully – relinquish our certainties.

It is striking how like a dream Leopold's story is. In his *Interpretation of Dreams*, Freud says that dreams are characterized by a quality he calls 'das Unheimliche' or 'the uncanny' – that is, when the familiarity of certain objects, persons or situations is interrupted or associated with other elements that are strange. In this story, the transformations occur quickly, as the young man who grew up in Iowa peers across the alien beauty of the New Mexican highlands. First, he sees a doe swimming through a rushing river – a dramatic image in its own right – which changes to an image of a she-wolf, who is almost immediately joined by a romping melee of six other wolves. Then, with blood on the slick rock, only the original wolf remains, and she dies before his eyes.

A dream can illuminate the innermost issues of our lives. Conversely, our cathartic experiences can become dreams, to which we return for clarification as we traverse the threshold of consciousness throughout the following nights and years. Sometimes events like the one Leopold describes here have an inadvertent quality – we stumble upon new understandings, often through our own error. He writes that '[i]n those days we had never heard of passing up a chance to kill a wolf'; Leopold only begins to wonder what his actions mean when both the other wolves and the other men have disappeared from the story, and he and the she-wolf are alone together as the green fire dies.

This story can be understood as a conversion experience reminiscent of St Paul on the road to Damascus, who was struck blind by a light that he had not been seeking, and staggered to his feet as a new man. Leopold, too, walked away from his experience changed. He is known today as the author of *A Sand County Almanac*, which has proven one of the most influential books in the American environmental movement. In the decades between the episode described in 'Thinking Like a Mountain' and the appearance of his book, Leopold was also a professor of natural resources and game management at the University of Wisconsin, and in this capacity he became one of the first and most powerful advocates of the importance of predators within a healthy ecosystem. Today, when wolves grace calendars, Christmas cards and the labels of micro-brews, it is hard to remember what a revolution there has been since the days when Leopold was so quick to kill one. We now know that the health of the trees and other vegetation on a mountain depends in part on wolves, which keep deer populations in check. But this far-reaching insight originated in visceral experience, and is most powerfully conveyed by a story. Thus, literature and science both contribute to thinking like a mountain.

Leopold's narrative, like Snyder's poem, illustrates how losses can be experienced as opportunities for deepened vision. Reversing the upward trends in population and consumption reflected in the graph of Snyder's poem will certainly require enormous efforts, but there is no other way out, and the question is whether we muster up the courage to relinquish our misperceptions and learn from them, or simply crash headlong and fall. While it is already impossible to avert some forms of environmental degradation, we can at least try to understand that these events are integral to our situation, and that the basis for making the necessary transformations must begin within.

The term 'crisis' is therefore preferable to 'problems' for describing our environmental situation, as it implies not just difficulty but a climax and juncture after which there will be notable change. The Japanese kanji for 'crisis' is a compound character, *kiki*, in which the radical for 'danger' stands beside that for 'opportunity'. Crises are critical turning points. Leopold would not have begun thinking like a mountain if he had not encountered, through his own heedlessness, the damage caused by the eradication of wolves. Without looking up now into our injured sky, we could not seriously contemplate altering our wasteful and unsustainable ways. Another venture into etymology, this time in English, reveals that the word 'sacrifice' is derived from the Latin words for 'holy' and 'to make'. The blood sacrifices involving ritual slaughter of animals practiced in ancient traditions around the world, from Israel to Greece to China, transformed the people who make them, allowing them to enter into new relationships with their gods.

Animals have always been killed, whether for meat or simply for the excitement of the chase. In the case of sacrifice, the difference is one of intent. And as Leopold's passage shows, the intention can emerge after the act itself, giving the act a meaning that it did not originally have, and making it a catalyst that generates a new stage in the growth of an individual or a community. In his essay

'The Land Ethic', Leopold surveys Western tradition, from Homer through the Bible, interpreting it as an ever-widening ethical circle: over time, rights are extended to women, slavery comes to be understood as morally abhorrent, and democracy emerges to protect individuals within the massive modern state. Today we are called upon to develop an ethic that includes the 'land and . . . the animals and plants which grow upon it'. Such an ethical extension 'is, if I read the evidence correctly, an evolutionary possibility and an ecological necessity'. Necessity is, in fact, our opportunity. If there were an alternative to dealing with the daunting problem of the momentum of human population growth and the environmental crisis that is arising from it, we would certainly pursue it. But nothing can substitute for the value of a livable earth, so we must make a start. And, in so doing, we may soon find ourselves living in a replenished and recharged world.

Admitting that the present course is no longer tenable can engender a sense of chastened humility, but this humility can also be a source of positive energy, leading to a heightened sense of community. Rather than laying blame for our current crisis on Western civilization, as some environmental thinkers tend to do, this crisis should inspire a careful rethinking of that tradition, placing special value on individuals who, in their own day, addressed the same types of ethical and psychological issues that confront us today. Ezekiel and Isaiah warned against mistreating the land; Jesus taught about the Kingdom of Heaven through stories of trees and vines; St Francis included the sun and moon, birds and wolves in his spiritual family – all of them are revealed to us in new ways by the mandates of the current crisis.

Environmental crisis also promotes deeper forms of cultural dialogue. The inclusion of courses in non-western languages and cultures at colleges like Middlebury, and an increasing emphasis on interdisciplinary studies, is part of this broadening conversation. Japan's nature literature is one source of revelation. For example, understanding the similarities between a Japanese author like Basho, and Wordsworth and Thoreau in the Western tradition, can bring new understandings to the familiar work of these writers, generating an awareness of their writings as visions and as dreams. This *haiku* by Basho speaks to the theme of gain through loss:

> Haru no hi ya
> mizu sae areba
> kurenokori.

A spring day–
If there are puddles on the ground,
the light may last a little longer.

Looking down at the sodden ground, one sees the sky. Even as the sun sets, it shines around us, gleaming at our very feet. Once again, contemplation of the atmosphere leads us back to earth. It also reverses our cultural momentum in another important way. One of the principal values of Japanese literature is its microcosmic perspective, the recognition that one can appreciate nature's beauty and connection with the human spirit not only amidst the sublime wilderness, but also in the local and undramatic way a puddle catches the light. This recognition is both stabilizing and consoling. Reading Basho reminds one that such little moments are important in a work like *Walden*, too. Basho frees us from being tied to an understanding of Thoreau simply as the representative of a particular political theological, or literary history. Thus, the two authors' affinity makes both of them freshly available and, in a surprising way, more contemporary.

This era of environmental crisis also promotes fuller dialogue between the humanities and the natural sciences. Just as we now discover common ground with non-Western cultures, so too do we remember that the achievements and insights of our culture grew out of, and need to be tested against, the dynamics of our living planet. Despite C.P. Snow's famous formulation about the division between scientific and humanistic education as 'the Two Cultures', liberal arts colleges need to expose all of their students to both realms of knowledge in a significant way. This is not easy, simply because there is so much more to know with each passing year. Yet unless educated people can respond to the present crisis with a wide range of information and broad powers of comprehension, we will never muster the clarity and energy necessary to survive.

To generate this capability to respond, it may be necessary – though painful – to relinquish our disciplinary and departmental categories and ask what skills and knowledge are truly essential, while paring away others. To integrate our curriculum in this way we need to locate cases, texts and problems that gather disparate disciplines within one educational circle. One of the many valuable aspects of Leopold's *Sand County Almanac* is that it can be discussed with equal value in classes taught by natural scientists, by specialists in public policy, and by literary critics. The book of *Genesis* and Darwin's *Origin of Species* might also be useful, cross-disciplinary keystone texts.

This educational synthesis sounds like a lot of work, but the merger of structures or ideas that had been held to be separate can also be energizing. It can be a strenuous joy to pursue a serious conversation about teaching, learning and citizenship in the context of environmental studies with colleagues from the natural sciences and the social sciences, as well as from other areas of the humanities. Such connections can, for example, lead to teaching methods that get students back out under the sky. For example, over the last several summers at Middlebury College's Bread Loaf School of English, students and instructors have rejoiced in courses that involve several weeks of hiking and camping in the mountains. These shared adventures heighten the experience of intellectual community, as participants read and write, walk the ridges with geologists, study the alpine flora with botanists, and share findings each evening in discussions around the campfire. It may also be possible to integrate such models more fully into teaching during the regular academic year.

Education needs to move beyond compartmentalization; the disciplines need to hit the trail together and share their stories out-of-doors. Another resource that can be brought to this task is the continent's native cultural traditions rooted in the land. In her essay 'Landscape, History, and the Pueblo Imagination', the Laguna novelist Leslie Marmon Silko has written about her own people's inclusive vision of the earth:

> The bare vastness of the Hopi landscape emphasizes the visual impact of every plant, every rock, every arroyo. Nothing is overlooked or taken for granted. Each ant, each lizard, each lark is imbued with great value simply because the creature is there, simply because the creature is alive in a place where any life at all is precious. Stand on the mesa edge at Walpai and look west over the bare distances toward the pale blue outlines of the San Francisco peaks where the ka-tsina spirits reside. So little lies between you and the sky. So little lies between you and the earth. One look and you know that simply to survive is a great triumph, that every possible resource is needed, every possible ally – even the most humble insect or reptile. You realize you will be speaking with all of them if you intend to last out the year.[11]

To live in such a landscape is to experience the equivalent of a perpetual environmental crisis. Mindfulness in practice and an enhanced sense of community can both be derived from the desert's starkness. Such austere circumstances place a premium on remembering the ancient stories of one's culture, on humility about our human prerogatives, on attentiveness to creatures that share the landscape with us, and on the necessity to curb our appetites and guard against wastefulness. They have long reminded the Laguna Pueblo to 'stay together/learn the flowers/go light', and the crisis into which the rest of humanity has now entered should prompt us to do the same.

Notes

1. A shorter version of this essay appeared under the title of 'Climbing the Crests' in the *Associated Departments of English Bulletin*, September, 1996.
2. G. Snyder, 'For the Children', *Turtle Island* (New York: New Directions, 1974), 86, reprinted by permission of the author.
3. D. Pimental *et al.*, 'Natural Resources and an Optimum Human Population', *Population and Environment* 15 (May 1994): 349.
4. Population Reference Bureau, *World Population Data Sheet* (Washington, DC: Population Reference Bureau, 1991), cited in Pimentel *et al.*, *ibid.*
5. R. Gelbspan, 'The Heat is On: The Warming of the World's Climate Sparks a Blaze of Denial', *Harper's* (December 1995).
6. B. McKibben, *The End of Nature* (New York: Random House, 1989).
7. E. Hoagland, 'Edward Abbey: Standing Tall in the Desert', *New York Times Book Review* (7 May 1989): 44.
8. L. White, Jr., 'The Historical Roots of Our Ecologic Crisis', reprinted in *Machina ex Deo* (Cambridge: MIT Press, 1988): 75–94.
9. W. Wordsworth, fragments from Alfoxden Notebooks, in *The Essential Wordsworth*, ed. Seamus Heaney (The Ecco Press, 1988): 33.
10. A. Leopold, 'Thinking Like a Mountain', in *A Sand County Almanac* (Oxford, 1989), 129–30.
11. L.M. Silko, 'Landscape, History and the Pueblo Imagination', in *Norton Book of Nature Writing*, ed. R. Finch and J. Elder (New York: W.W. Norton, 1991): 894.

18 Ecological Security: Micro-Threats to Human Well-Being

Dennis Pirages

The decade between the end of the Cold War and the beginning of the next millennium has been marked by transitions in perceptions about threats to human security. Missiles are being dismantled in Siberia and the American midwest, and the nuclear threat to human survival is much diminished. Although Iraq still covets Kuwait, and other disputes over territory also simmer in the Middle East, for the most part, large-scale, high-intensity warfare across borders is on the wane. But these traditional kinds of insecurities are quickly being replaced by other threats arising from demographic shifts and environmental change.

For most of recorded history the focus of security policies has been on military threats to human well-being. Given the destruction that has been associated with military conflicts since ancient times, it is understandable that human security has been defined primarily in military terms. Threats from other peoples have been readily understood, and defenses could be mounted against them. While other types of security challenges such as famines, plagues and pestilence may have been much more destructive, understanding of these matters has been lacking, and remedies have not been readily apparent. In recent decades, the experience of two world wars, the Korean and Vietnamese wars, and the nuclear stalemate of the Cold War has helped to firmly establish departments of defense in the business of building military forces in an effort to deter potential human adversaries.

The expressed purposes of defense spending have traditionally been preservation of the state, protection of the national interest, and prevention of large-scale death and destruction. Given *Homo Sapiens'* long experience of conflict, the primary focus of collective defense efforts has been prevention of civil disorders, cross-border incursions, and the associated civilian and military casualties. However, while military dangers still exist, the locus of conflict is now shifting from military to environmental matters, and from international to intranational quarrels in ethnically divided and sometimes 'failing' states.[1] As a result, in recent years US military forces have been involved in refereeing disputes among clans in Somalia, providing relief to victims of conflict in Bosnia, restoring some semblance of order to ecologically and politically ravaged Haiti, and providing disaster relief to flood victims in Bangladesh.

This chapter reassesses threats to human security based on the notion that the function of security policy is protection of the state, prevention of large-scale loss of human life, and preservation of human well-being. But it argues that the focus of security-thinking and efforts should be broadened to encompass more than just military challenges. Many of these 'new' challenges are not really new, but the ability to deal with them is much greater now than in the past. Traditionally they have been regarded as the work of God or gods, and thus beyond human remedy, so there was little incentive to formulate security policies to deal with them. Thus, the Black Death (bubonic plague) carried by Rattus rattus in the fourteenth century was certainly perceived by the people affected as a major challenge to their well-being – nearly 40 per cent of those exposed perished – but the only defensive actions that were thought to be efficacious at that time were prayer and self-flagellation.[2] Today, however, advances in medical science and in environmental understanding are opening up new opportunities to make the world a more secure place.

In the twentieth century, as in the past, casualties from military engagements actually pale in comparison with the human suffering and premature deaths caused by famine and disease. While the eyes of the world focused on the military horrors of World War I, an influenza virus spread around the world from the United States, exacting more than 20 million casualties, many times the total number of battlefield deaths.[3] Likewise, in recent decades far fewer deaths are associated with civil strife and international conflict each year than the approximately 20 million who die annually from infectious diseases. In fact, the actual ratio between premature deaths in military engagements and those due to other causes may not be so different now compared with previous centuries – the latter has always been the much larger number. However, today our ability to prevent, understand and deal with health and environmental problems is much enhanced if the appropriate resources are made available.

ECOLOGICAL SECURITY

Homo Sapiens is one species among millions, albeit a very significant one. Like other species, human beings live in basic population units. Populations are defined as 'dynamic systems of individuals . . . that are potentially capable of interbreeding with each other'.[4] Human populations are generally demarcated by a common language and culture. The thousands of human populations that once roamed the earth's surface have been consolidated into an ever-changing state system only relatively recently. Sometimes human populations are fairly contiguous with state borders, but at other times they are not.

Homo Sapiens has evolved within the changing constraints of a physical environment shared among human populations, as well as with various kinds of microorganisms and populations of other species. This coevolutionary process has been influenced by environmental shifts, human migrations, changes in human behaviour, the growth and decline of populations of other species, and the changing fortunes of micro-organisms.

An ecological approach to defining human security moves beyond environmentalism and focuses on the evolving relationships critical to the future welfare of the human race, especially the four relationships between human populations and:

1. their physical environment;
2. other human populations;
3. other species; and
4. pathogenic micro-organisms.

Ecological security is predicated upon maintaining delicate equilibriums in each of these four relationships.

First, ecological security is enhanced when the resource demands of human populations can be met within the physical limits of the territory that they occupy. Human populations, like those of other species, tend to expand until they reach the carrying capacity of their territory. Prior to the twentieth century, deaths from famine and environmental pollution that occasionally accompanied excessive population growth were largely localized. But in this century, population growth, combined with the demands of industrialization, has begun to stress environmental services on a global scale. The population explosion in the South and the increasing environmental demands of industrialization are already running up against the carrying capacity of the land there, leading to environmental degradation, increased vulnerability to disease, and frequently to violent conflict.[5]

Secondly, ecological security is also enhanced by maintaining equilibrium among human populations. The rapid growth of human populations in the face of resource endowments that are inadequate to meet their demands has frequently generated pressures to expand territory by moving against neighbours.[6] Thus, war among nations, the focus of more traditional security concerns, is still readily incorporated into the broader ecological security framework proposed here. Both cross-border military incursions as well as ethnic turmoil are often manifestations of differentially growing human populations contesting for access to territory and resources. For example, persisting tensions between Iraq and Iran, or Israel and its Arab neighbours, are certainly exacerbated by very high rates of population growth in the region.[7] And various regional skirmishes, including those involving the Kurds in the Middle East or various tribes in parts of Africa, are essentially turf battles exacerbated by growing populations.

The third dimension of ecological security involves preserving equilibrium between human populations and competing species. Human beings share the global habitat with significant numbers of other large species, and often find themselves directly competing with them for resources. Thus, the periodic rampages by elephants and tigers in areas recently settled by humans in India, or the nuisance activities of bears and deer in suburban areas of the US, are examples of conflicts between man and beast for territory and resources. The Biblical plagues of locusts, as well as the contemporary battle with the California medfly, reflect an even more pronounced threat to security: direct competition for shared food supplies. While the inter-species competition arising from the unprecedented expansion of human numbers more often leads to the extinction of other species than to human suffering, such trends also pose a potential indirect threat to human well-being because of the unknown impacts of the considerable loss of genetic diversity.[8]

Finally, and perhaps most importantly, ecological security is enhanced by preserving the delicate equilibrium that has, for the most part, persisted between the human body and pathogenic microorganisms. *Homo Sapiens* has coevolved with a wide variety of microorganisms, and this relationship has generally been marked by peaceful coexistence, resulting from immunities developed as a result of human exposure to these pathogenic microbes.[9] Occasionally, however, this peace has been shattered by outbreaks of disease. Rats and their passengers, travelling the trade routes from China, were a key vector for the outbreak of bubonic plague that struck down many previously unexposed Europeans during the Middle Ages. Similarly, the 'discovery' of the Americas opened up a two-way flow of microorganisms that wreaked havoc in both Europe and the Americas.[10] For many reasons, human populations are coming into contact with new pathogenic micro-organisms as well as developing new vulnerabilities to old diseases, and there is a general breakdown in immunity.[11] The next century may well see a resurgence of plagues brought about by changes in man–microbe relationships.

UPSETTING THE EQUILIBRIUM

The rapid growth of human populations has heightened the general sense of ecological insecurity reflected in the deterioration of the ozone layer, global warming, other types of environmental degradation, famine and various forms of conflict. But of the various potential threats to ecological security, one of the most serious is the growing disequilibrium between *Homo Sapiens* and pathogenic micro-organisms. Threats from the host of microorganisms with which human populations have coevolved are certainly not new. History is littered with the remains of societies that have succumbed to various diseases. But there is considerable evidence that large-scale demographic changes shifts in patterns of human behaviour, technological changes, and increasing interdependence and globalization are making *Homo Sapiens* much more vulnerable to new and resurgent diseases. 'At the root of the resurgence of old infectious diseases is an evolutionary paradox: the more vigorously we have assailed the world of microorganisms, the more varied the repertoire of bacterial and viral strains thrown up against us'.[12]

In the US there have recently been several bouts with pathogenic microbes, including an outbreak of cryptosporidium that killed 104 people in Milwaukee, a lethal episode of hantavirus in the four corners area of the southwest, several widespread cases of food poisoning, influenza epidemics, an outbreak of antibiotic-resistant intestinal disease in New York, attacks by 'flesh eating' bacteria, and the appearance of drug-resistant bacteria in many day-care centres and hospitals, all unexpected in either scope or novelty.

In other parts of the world the disease situation is much more grim. It is estimated that more than 500 million people are suffering from tropical diseases. Approximately 400 million people are suffering from malaria, 200 million from schistosomiasis, and 100 million from lymphatic filariasis.[13] In addition, the rapid spread of the AIDS virus has created a crisis situation in countries as diverse as Thailand, India and South Africa. It is estimated that by the year 2000, fully one-fifth of the adult population of South Africa will be HIV positive.[14] In India, although not nearly as large a proportion of the population will be affected, because of its huge size nearly six million people are expected to be HIV positive by that time.[15] In Latin America, more than a million people have been struck with cholera in recent years, and several thousand have died.[16] An outbreak of pneumonic plague in India recently killed hundreds of people and disrupted air travel in the region.[17] And the sporadic appearance of deadly ebola-type viruses in several tropical locations has created near-crisis situations.

It is somewhat ironic that the latter half of the twentieth century has given birth to a biomedical revolution, yet at the same time many traditional diseases are making a comeback and new microbes are on the attack. New discoveries have seemingly given *Homo Sapiens* additional weapons in the struggle against infectious diseases, but changes in human behaviour and in the human condition are opening up new opportunities for pathogenic micro-organisms. There are at least five significant changes in human behaviour, technology and the physical environment that are behind this microbial resurgence: large-scale demographic transformations; increasing poverty worldwide; growing permissiveness in human behaviour; the negative aspects of technological innovation; and environmental change. These changes are all responsible for upsetting the existing equilibrium with pathogenic micro-organisms.

DEMOGRAPHIC DISCONTINUITIES

There are at least four kinds of demographic changes that have epidemiological significance: rapid population growth, urbanization, migration and aging. The world is divided demographically between the slower growth North and the still rapidly growing South, and although it is now past the period of most rapid population growth, there is still tremendous demographic momentum pent up in the young populations in the South. As a result, the current world population of 5.8 billion is projected to grow to 8.2 billion by the year 2025.[18] At that time, 6.9 billion people will reside in what is now the less-industrialized world. A large portion of these people will be pressing into densely populated 'megacities', where disease organisms can spread rapidly. By the year 2034, for example, Mexico City and Shanghai could have populations of 39 million, Beijing 35 million, Sao Paulo 32 million, and Bombay 31 million.[19] As ever larger numbers of people wind up living in squalid conditions in densely populated cities, the opportunities for viruses, bacteria and parasites to prey on human beings increase.

Rapid population growth in the less-industrialized countries is also generating pressures for people to venture into remote and previously isolated areas. In both Latin America and Africa, migrants are clearing land on the edges of previously isolated tropical rainforests. These trespassers are venturing into territories that harbour large numbers of unknown micro-organisms, many of which are potentially pathogenic. As the forests continue to fall before the ax and the plow, these micro-organisms have a much better opportunity to move into human populations that have little resistance to them.

The recent appearances of the ebola virus among people living on the edge of the rainforest in Zaire, as well as outbreaks of hemorrhagic fever in forest areas of Bolivia, are evidence of these dynamics. Fortunately, extremely potent viruses such as ebola have obvious symptoms and kill their hosts within only a few days, and it has been possible to isolate victims and keep the virus from spreading to large numbers of people. Future viruses might not be so obvious, and could spread rapidly before being detected.

Large-scale population movements, often the result of ethnic conflicts, political repression, and military violence, can be another factor in the spread of diseases. Thus, in countries like Rwanda, Zaire and the former Yugoslavia, poverty-stricken refugees living in squalid conditions in refugee camps or on the move to avoid massacres often fall prey to infectious diseases, and they carry these with them when they eventually leave the camps as well.

Somewhat paradoxically, even the demographic changes associated with approaching zero population growth in the industrial countries can affect the delicate man–microbe equilibrium. Low birth rates, combined with medical innovations that prolong human life, are creating greyer populations. The OECD estimates that by the year 2030, 20 per cent of the populations of the US and Japan, and more than 25 per cent of people in Switzerland and Germany, will be over 65.[20] Greying populations are more vulnerable to communicable diseases and chronic illnesses, and require much more extensive medical care. It is estimated that in the US, chronic illnesses now consume about $425 billion annually in medical costs. This is projected to rise (in 1990 dollars) to $798 billion by the year 2030.[21]

GROWING WORLD POVERTY

Rapid population growth on the south side of the global demographic divide has been accompanied, in many countries, by deteriorating economic circumstances. Between 1980 and 1993, GNP per capita declined annually in 52 countries, or 40 per cent of those for which the World Bank keeps relevant statistics.[22] In countries ranging from major oil exporters to those that have historically been economically troubled, this traditional measure of industrial progress has indicated an alarming decline in living standards. Among the most significant of the economic laggards were countries as diverse as Georgia (with an average annual decline in GNP of 6.6 per cent), Nicaragua (-5.7 per cent), Côte d'Ivoire (-4.6 per cent), and the United Arab Emirates (-4.4 per cent). Collectively, the low income countries, excluding China and India, showed almost no increase in per capita GNP, the lower-middle-income economies saw a 0.5 per cent annual deterioration, sub-Saharan Africa witnessed a 0.8 per cent annual deterioration, the Middle East and North Africa a 2.4 per cent annual decline, and the middle-income countries of Europe and Central Asia saw a 0.3 per cent decline annually. Even the industrial countries experienced relatively slow growth during the period, particularly when compared with historical standards.

The causes of this increasing impoverishment are varied, but it is clear that the spread of the industrial revolution and its assumed benefits to many parts of the world can no longer be guaranteed. More to the point, however, is the impact of poverty on human health prospects. While annual per capita health care expenditures amount to nearly $3000 in the US, in Zaire, Tanzania, Sierra Leone and Mozambique the rate is five dollars or less per person.[23] The recent and rapid impoverishment of the former Soviet Union has also been associated with a significant increase in the incidence of various diseases.[24] Should the ebb in the development tide not be reversed, innovations in medicine and biomedical technology may have little impact on the afflicted in countries that do not have sufficient financial resources to devote to health care.

INCREASING HUMAN PERMISSIVENESS

From an evolutionary perspective, changes in the ways in which *Homo Sapiens* lives can also have significant impacts on ecological security. Historically, major changes in human organization and behaviour have had serious epidemiological consequences – the squalid, densely-packed medieval cities, for example, offered a remarkably good environment for the spread of bubonic plague. Today, large-scale urbanization, particularly in the less-industrialized parts of the world, is similarly creating squalid and unmanageable slums in which diseases can rapidly spread among the victims of urban poverty.

Similarly, almost every period of sexual revolution in history has had significant health consequences for those breaking free of traditional behavioural constraints.[25] Value and behavioural changes accompanying the industrial revolution have created permissive societies in which large numbers of people have now strayed from ecologically acceptable behaviour patterns. Over the last three decades, a worldwide sexual revolution has led to the proliferation of sexually transmitted diseases such as herpes, syphilis and gonorrhea, and has facilitated the spread of AIDS. Much more widespread use of illicit drugs, particularly those injected with shared needles, is another innovation in human behaviour that is tipping the balance between *Homo Sapiens* and micro-organisms. Such changes in behavioural patterns on a global scale have no precedent. The worldwide breakdown in morality and the related urban crime wave is also overcrowding prisons, where many prisoners have HIV-weakened immune systems that cannot fight off tuberculosis. This contributes to the spread of an old disease that is developing resistance to most antibiotics.[26]

THE DARK SIDE OF TECHNOLOGY

Technological innovation is another factor influencing the balance between human populations and micro-organisms. Progress in biomedical technology has unquestionably created a new armoury of techniques with which to do battle with pathogenic microorganisms. Other technological innovations have also increased disease resistance among human beings. But technological innovations often have unforeseen consequences, some of which have a negative impact on ecological security.

For example, transportation innovations have increased the speed with which ever-larger numbers of people – and their accompanying microbes – move, and the territory over which they range. People and products are in constant motion in the emerging global village. Thus, diseases such as the 'Wuhan' flu or the 'Seoul' virus may originate in Asia, but have their harshest impact on populations thousands of miles away that have much less immunity to these viruses. The large-scale and rapid movement of people and products internationally has given birth to a host of global hitchhikers – viruses, bacteria and pests – that sneak rides into new environments, where they often flourish. Thus the Seoul virus (hemorrhagic fever) has appeared in Baltimore, Maryland, apparently having been transported from Korea by wharf rats.

Larger migrant organisms have also been transforming the environment and adding to ecological insecurity. Nomadic zebra mussels, for example, have invaded the US, having travelled from Europe in the ballast water of cargo ships. The mussels are flourishing in their new environment, and have done hundreds of millions of dollars in damage to aquatic ecosystems and water intakes, particularly in the Great Lakes region.[27] Dozens of other such migrants, ranging from killer bees to California's superbugs (*Bemisia tabaci*), have flourished after being transported into new, previously unexploited environments.

New methods of producing and distributing food products can also adversely impact health security. Agricultural produce now comes from megafarms, is processed in food factories, distributed in megamarkets and consumed in fast-food emporiums. Not surprisingly, there have been several recent, large-scale outbreaks of food-related disease in industrial countries. Contaminated hamburgers in the northwestern US, and ice cream from the midwest, have been responsible for debilitating outbreaks of illness. 'Mad cow' disease has devastated herds in the UK and possibly caused illness in human beings, resulting in sharply reduced consumption of beef and a halt in beef exports from the country. In Japan, an outbreak of *e-coli* bacteria felled thousands of school children. The more centralized and large-scale production and distribution of food thus raises the spectre of potentially more severe outbreaks of disease rapidly spreading through human populations. Even the apparently beneficial use of antibiotics can have troubling outcomes. Mutation of bacteria and viruses is part of the evolutionary process, but it allows these organisms to adapt to the large-scale application of antibiotics and other drugs, rapidly reshaping the microbial world in both positive and negative ways. Since there are few ecologically-based limits on the use of these chemicals, their widespread use on humans and farm animals has become a major factor adversely affecting the human–microbe equilibrium. Continued exposure to these chemicals has accelerated mutation processes and transformed many micro-organisms into drug-resistant forms. For example, a resistant strain of gonorrhea first emerged in the Philippines more than a decade ago. It has now moved into many new environments, and there is only one drug left that can successfully treat it. A growing number of resistant micro-organisms have emerged from encounters with antibiotics and other drugs, resulting in much less treatable forms of tuberculosis, pneumonia, dysentery and other diseases.[28]

ENVIRONMENTAL CHANGES

Transformation of existing environments, like the transfer of organisms into new zones, can also upset the delicate equilibrium between *Homo Sapiens* and pathogens. For example, a fatal outbreak of hantavirus in the southwestern US was probably triggered by changes in rainfall patterns that increased food availability for rodents, thereby facilitating the growth of the rodent populations that carry the virus.[29] A more obvious example is the migration of killer bees into parts of the US that were previously thought to be too cold for them to survive. Warmer summers and winters have apparently extended their range.

There are three readily identifiable types of environmental changes that could have some effect on the future evolution and spread of serious diseases: the multifaceted impacts of global warming, the breakdown of the ozone layer and the associated increased penetration of ultraviolet radiation, and the accidental creation of new kinds of polluted primordial soups within which various microbes can mingle.

Projecting the future course of global warming is an inexact science, and numerous estimates of likely temperature increases have been made with no general agreement reached. But even small global temperature increases are likely to have a significant impact on regional temperature and rainfall patterns. This, in turn, could lead microbe-carrying animals to range into new territories, and tropical diseases such as yellow fever could make inroads into formerly temperate regions.[30] An increase in ultraviolet radiation, meanwhile, may lead directly to increasing cases of melanoma and cataracts, and may also have indirect effects by suppressing the human immune system.[31] Finally, the waste products of human activities, ranging from sewage to chemicals, increasingly mingle in rivers, bays, streams and coastal estuaries, where mutations and recombination can rapidly take place.[32]

BUILDING ECOLOGICAL SECURITY

Creating a more secure world for the next millennium requires restoration of the four equilibriums that define ecological security. This project will require a paradigm shift with respect to the conceptualization, research and funding of security theory and policy formulation. Some preliminary suggestions to begin restoring equilibrium with micro-organisms are offered here.

The task of rebuilding microsecurity can move forward on two levels. First, a major change in approaches to developing and funding security policy is needed to deal directly with the emerging threat of new and resurgent diseases. However, this is not easily accomplished in an era of perceived scarcities, when the US public health infrastructure has been repeatedly downsized, rather than expanded. Second the broader problem of sociocultural drift that has been a significant factor worldwide in destabilizing the human–microbe relationship must be addressed.

Some timid steps have been taken to deal with potential future outbreaks of serious disease. Advisory committees have met in the US, and some cosmetic changes have been made in information-gathering practices. But no significant new policies or funding have been put into place to deal with immediate threats. As a minimum, a realistic microsecurity policy should include a much more vigorous surveillance and reporting network to gauge the threat represented by newly emerging and resurgent diseases. At present, international monitoring and response capabilities are poorly developed and largely limited to the US Centers for Disease Control and Prevention and the World Health Organization (WHO), both organizations that are now woefully under-funded. The WHO now requires governments only to report outbreaks of cholera, plague and yellow fever, and many governments try to cover up disease outbreaks that do occur. An early warning network is now being set up as a joint US–European Union venture, but it will only cover diseases in those regions and others that happen to come to the attention of scientists.

A major cooperative international effort is required to monitor and deal with potentially lethal microorganisms. Deaths and impairments from serious diseases should be considered as much a threat to state and human security as battlefield encounters. But most important, in an era of public-sector downsizing, it is critical to increase funding for monitoring and treating infectious diseases. Thinking strategically, tremendous worldwide improvements in the human condition could be made by foregoing the production of just one stealth bomber.

In the long-term, however, the four ecological equilibriums can only be restored and maintained by basing security policies on a better understanding of evolutionary relationships. This may require the equivalent of a sociocultural genome project, an analysis of what human behavioural and value-changes have done to destabilize these critical relationships. But this is much more difficult to accomplish than using new antibiotics or other technological fixes as temporary repairs, largely because such a genome project would require a type of objective analysis of changing human values and behaviour that is actively discouraged in today's permissive, *laissez-faire* world.

Notes

1. See Gerald B. Helman and Steven R. Ratner, 'Saving Failed States', *Foreign Policy* 89 (Winter 1992–93).
2. See William H. McNeil, *Plagues and Peoples* (Garden City, New York:

Anchor Press/Doubleday, 1976), Chapter IV.
3. See Alfred Crosby, *America's Forgotten Pandemic: The Influenza of 1918* (Cambridge: Cambridge University Press, 1989).
4. Kenneth Watt, *Principles of Environmental Science* (New York: McGraw-Hill, 1973), 1.
5. This has been documented in Thomas F. Homer-Dixon, 'On the Threshold: Environmental Changes as Causes of Acute Conflict', *International Security* (Fall 1991); and Thomas F. Homer-Dixon, 'Environmental Scarcities and Violent Conflict: Evidence from Cases', *International Security* (Summer 1994).
6. This theme is developed in Nazli Choucri and Robert North, *Nations in Conflict* (San Francisco: W. H. Freeman, 1975).
7. John R. Weeks, 'The Demography of Islamic Nations', *Population Bulletin* (December 1988).
8. See Paul Ehrlich and Anne Ehrlich, *Extinction: The Causes and Consequences of the Disappearance of Species* (New York: Random House, 1981).
9. See Marc Lappe, *Evolutionary Medicine: Rethinking the Origins of Disease* (San Francisco: Sierra Club Books, 1994).
10. See McNeil, note 2 above, Chapters III, IV and V.
11. See the essays in Stephen S. Morse, ed., *Emerging Viruses* (New York: Oxford University Press, 1993).
12. Lappe, 8, note 9 above. See also, Laurie Garrett, *The Coming Plague: New Emerging Diseases in a World Out of Balance* (New York: Farrar, Straus & Giroux, 1994).
13. World Health Organization, *The World Health Report 1995* (Geneva: World Health Organization, 1995), Table I.
14. Lynne Duke, 'Opening of South Africa Brings Rapid Advance of AIDS', *The Washington Post* (23 July 1995).
15. John Ward Anderson, 'India Seen as Ground Zero in Spread of AIDS to Asia', *The Washington Post* (17 August 1995).
16. Don Podesta, 'For South America a Return to a Time of Cholera', *The Washington Post* (28 February 1993).
17. Molly Moore and John Ward Anderson, 'Plague Turns India into Region's Pariah', *The Washington Post* (2 October 1994).
18. Figures are from Population Reference Bureau, *World Population Data Sheet 1996* (Washington, DC: Population Reference Bureau, 1996).
19. Figures are from Leon Bouvier, 'Planet Earth 1984–2034', *Population Bulletin* (February 1984).
20. OECD, *Aging Populations: The Social Policy Implications* (Paris: OECD, 1988), 22.
21. Data are from the World Bank, *World Development Report 1995* (New York: Oxford University Press, 1995), Table 1.
22. 'Chronic Illnesses Cost U.S. $425 Billion a Year', *The Washington Post* (13 November 1996).
23. Data from World Health Organization, *The World Health Report 1995* (Geneva: World Health Organization, 1995), Table A3.
24. John Maurice, 'Russian Chaos Breeds Diphtheria Outbreak', *Science* 267 (10 March 1995).

298   *Ecological Security*

25. See Theodore Rosebury, *Microbes and Morals: The Strange Story of Venereal Disease* (New York: Viking Press, 1971).
26. Barry R. Bloom and Christopher J.L. Murray, 'Tuberculosis: Commentary on a Reemergent Killer', *Science* 275 (21 August 1992).
27. 'Zebra Mussel Invasion Threatens U.S. Waters', *Science* 249 (21 September 1990). See also 'Biological Immigrants Under Fire', *Science* 254 (6 December 1991).
28. See Else Taenia 'Many Infectious Diseases Thwart Drugs', *The Wall Street Journal* (21 August 1992).
29. James M. Hushes *et al.*, 'Hantavirus Pulmonary Syndrome: An Emerging Infectious Disease', *Science* 262 (5 November 1993).
30. See Fillip H. Martin and Miriam G. Lefebre, 'Malaria and Climate: Sensitivity of Malaria Potential Transmission to Climate', *Ambio* 24 (June 1995).
31. See 'Environmental Effects of Ozone Depletion', special issue of *Ambio* XXIV (May 1995).
32. See Garrett, 557–68, note 12 above.

Index

308 *Index*